# PARALLEL ALGORITHM DERIVATION AND PROGRAM TRANSFORMATION

# THE KLUWER INTERNATIONAL SERIES IN ENGINEERING AND COMPUTER SCIENCE

## OFFICE OF NAVAL RESEARCH
Advanced Book Series

*Consulting Editor*

**André M. van Tilborg**

**Other titles in the series:**

**FOUNDATIONS OF KNOWLEDGE ACQUISITION: Cognitive Models of Complex Learning**, edited by Susan Chipman and Alan L. Meyrowitz
ISBN: 0-7923-9277-9

**FOUNDATIONS OF KNOWLEDGE ACQUISITION: Machine Learning**, edited by Alan L. Meyrowitz and Susan Chipman
ISBN: 0-7923-9278-7

**FOUNDATIONS OF REAL-TIME COMPUTING: Formal Specifications and Methods**, edited by André M. van Tilborg and Gary M. Koob
ISBN: 0-7923-9167-5

**FOUNDATIONS OF REAL-TIME COMPUTING: Scheduling and Resource Management**, edited by André M. van Tilborg and Gary M. Koob
ISBN: 0-7923-9166-7

# PARALLEL ALGORITHM DERIVATION AND PROGRAM TRANSFORMATION

*edited*
*by*

**Robert Paige**
*New York University*

**John Reif**
*Duke University*

**Ralph Wachter**
*Office of Naval Research*

**SPRINGER SCIENCE+BUSINESS MEDIA, LLC**

**Library of Congress Cataloging-in-Publication Data**

Parallel algorithm derivation and program transformation / edited by Robert Paige, John Reif, Ralph Wachter.
p. cm. -- (The Kluwer international series in engineering and computer science ; SECS 0231)
Includes bibliographical references and index.
ISBN 978-1-4757-8275-2 ISBN 978-0-585-27330-3 (eBook)
DOI 10.1007/978-0-585-27330-3
1. Parallel programming (Computer science) 2. Computer algorithms. I. Paige, Robert A. II. Reif, J. H. (John H.) III. Wachter, R. F. IV. Series
QA76.642.P35 1993 93-1687
CIP

---


Originally published by Kluwer Academic Publishers in 1993
Softcover reprint of the hardcover 1st edition 1993



*Printed on acid-free paper.*

# TABLE OF CONTENTS

## LIST OF ATTENDEES AT THE ONR WORKSHOP ON PARALLEL ALGORITHM DERIVATION AND PROGRAM TRANSFORMATION


**Sandeep Bhatt**
Computer Science Dept.
Yale University
New Haven, CT 06520
bhatt-sandeep@cs.yale.edu

**James M. Boyle**
Math & Computer Science Div.
Argonne National Laboratory
Building 221
9700 South Cass Ave.
Argonne, IL 60439-4844
boyle@mcs.anl.gov

**Jiazhen Cai**
Courant Institute
Computer Science Dept.
New York University
New York, New York 10012
cai@cs.nyu.edu

**David Callahan**
Tera Computer
david@tera.com

**Johnny Chang**
Courant Institute
Computer Science Dept.
New York University
New York, New York 10012
chang@cs.nyu.edu

**Thomas Cheatham**
Dept. of Computer Science
Harvard University
Cambridge, MA 02138
cheatham@harvard.harvard.edu

**Xushen Chen**
Courant Institute
Computer Science Dept.
New York University
New York, New York 10012

**Marina Chen**
Dept. of Computer Science
Yale University
P.O. Box 2158 Yale Station
New Haven, Conn. 06520
chen-marina@cs.yale.edu

**Young-il Choo**
Dept. of Computer Science
Yale University
P.O. Box 2158 Yale Station
New Haven, Conn. 06520
choo-young-il@cs.yale.edu

**Richard Cole**
Courant Institute
Computer Science Dept.
New York University
New York, New York 10012
cole@cs.nyu.edu

**Martin S. Feather**
USC/ISI
4676 Admiralty Way
Marina del Rey, CA 90291
feather@vaxa.isi.edu

**Apostolos Gerasoulis**
Dept. of Computer Science
Rutgers University
New Brunswick, NJ 08903
gerasoulis@cs.rutgers.edu

**Ben Goldberg**
Courant Institute
Computer Science Dept.
New York University
New York, New York 10012
goldberg@cs.nyu.edu

**Allen Goldberg**
Kestrel Institute
3620 Hillview Ave.
Palo Alto, CA 94304
goldberg@kestrel.edu

**Fritz Henglein**
DIKU
Universitetsparken 1
DK-2100 Kobenhavn 0
Denmark
henglein@diku.dk

**Donald B. Johnson**
Dept. of Math and CS
Dartmouth College
Hanover, NH 03755
djohnson@dartmouth.edu

**Neil Jones**
DIKU
Universitetsparken 1
DK-2100 Kobenhavn 0
Denmark
neil@diku.dk

**Zvi Kedem**
Courant Institute
Computer Science Dept.
New York University
New York, New York 10012
kedem@nyu.edu

**S. Rao Kosaraj**
Dept. of Computer Science
Johns Hopkins University
Baltimore, MD 21218
kosaraju@cs.jhu.edu

**Mike Langston**
Dept. of Computer Science
University of Tennessee
Knoxville, TN 37996
langston@cs.utk.edu

**David Lillie**
Dept. of Computing
Imperial College
Queens Gate
London SW7 2AZ, UK
djl@doc.ic.ac.uk

**Fillia Makedon**
Dept. of Math and CS
Dartmouth College
Hanover, NH 03755
makedon@dartmouth.edu

**Lambert Meertens**
Mathematisch Centrum
Kruislaan 413
1098 SJ Amsterdam
The Netherlands
lambert@cwi.nl

**Gary L. Miller**
Carnegie Mellon University
School of Computer Science
Schenley Park
Pittsburgh, PA 15213-3890
glmiller@cs.cmu.edu

**Bud Mishra**
Courant Institute
Computer Science Dept.
New York University
New York, New York 10012
mishra@cs.nyu.edu

**Bob Paige**
Courant Institute
Computer Science Dept.
New York University
New York, New York 10012
paige@cs.nyu.edu

**Krishna Palem**
IBM T. J. Watson Research Ctr.
P.O. Box 218
Yorktown Heights, N.Y. 10598
palem@cs.nyu.edu

**Helmut Partsch**
Dept. of Informatics
Catholic University of Nijmegen
NL-6525 ED Nijmegen, The Netherlands
helmut@cs.kun.nl

**Peter Pepper**
Tech. Universität Berlin
FB 20, Institut fuer Angewandte Informatik, Sekr. SWT FR 5-6
Franlinstrasse 28-29
D-1000 Berlin 10, Germany
pepper@cs.tu-berlin.de

**Alberto Pettorossi**
University of Rome II
c/o IASI CNR
Viale Manzoni 30
I-00185 Roma, Italy
adp@iasi.rm.cnr.it

**Jan Prins**
Computer Science Dept.
University of North Carolina
Chapel Hill, NC
prins@cs.unc.edu

**John Reif**
Computer Science Dept.
Duke University
Durham, N.C. 27706
reif@cs.duke.edu

**Larry Rudolph**
IBM T. J. Watson Research Ctr.
P.O. Box 218
Yorktown Heights, N.Y. 10598
rudolph@watson.ibm.com

**Jack Schwartz**
Courant Institute
Computer Science Dept.
New York University
New York, New York 10012
schwartz@cs.nyu.edu

**Doug Smith**
Kestrel Institute
3620 Hillview Ave.
Palo Alto, CA 94304
smith@kestrel.edu

**Rob Strom**
IBM T. J. Watson Research Ctr.
P.O. Box 218
Yorktown Heights, N.Y. 10598
strom@watson.ibm.com

**Carolyn Talcott**
Dept. of Computer Science
Stanford University
Stanford, CA 94305
talcott@cs.stanford.edu

**Thanasis Tsantilas**
Dept. of Computer Science
Columbia University
New York, New York 10027
thanasis@cs.columbia.edu

**Valentin Turchin**
CUNY
turcc@cunyvm.cuny.edu

**Uzi Vishkin**
University of Maryland
UMIACS, A. V.
Williams Building
College Park, MD 20742-3251
vishkin@umiacs.umd.edu

**David S. Wile**
Information Sciences Institute
4676 Admiralty Way
Marina del Rey, CA 90291
wile@vaxa.isi.edu

**Tao Yang**
Dept. of Computer Science
Rutgers University
New Brunswick, NJ 08903
yang@cs.rutgers.edu

**Allan Yang**
Dept. of Computer Science
Yale University
P.O. Box 2158 Yale Station
New Haven, Conn. 06520
yang-allan@cs.yale.edu

**Chee Yap**
Courant Institute
Computer Science Dept.
New York University
New York, New York 10012
yap@cs.nyu.edu

# SPEAKERS

| | |
|---|---|
| **Uzi Vishkin**<br>*University of Maryland &*<br>*University of Tel Aviv* | An Introduction to Parallel Algorithms |
| **Mike Langston**<br>*University of Tennessee* | Resource-Bounded Parallel Computation |
| **Lambert Meertens**<br>*Mathematisch Centrum* | Deriving Parallel Programs from Specifications |
| **Doug Smith**<br>*Kestrel Institute* | Automating the Design of Algorithms |
| **S. Rao Kosaraju**<br>*Johns Hopkins University* | Tree-based Parallel Algorithms |
| **Peter Pepper**<br>*Tech. Universitaet Berlin* | Deductive Derivation of Parallel Programs |
| **Alberto Pettorossi**<br>University of Rome II<br>IASI CNR | The Use of the Tupling Strategy in the Development of Parallel Programs |
| **Jan Prins**<br>*University of North Carolina* | Derivation of Efficient Bitonic Sorting Algorithms |
| **James M. Boyle**<br>*Argonne National Lab.* | Programming Models to Support Transformational Derivation of Parallel Programs |
| **Sandeep Bhatt**<br>*Yale University* | Mapping Computations onto Machines |
| **David Callahan**<br>*Tera Computer Co.* | Recognizing and Parallelizing Bounded Recurrences |
| **Marina Chen**<br>**Young-il Choo**<br>*Yale University* | Generating Parallel Programs from Algorithmic Specifications |
| **Neil Jones**<br>*DIKU* | Partial Evaluations |

| | |
|---|---|
| **Valentin Turchin**<br>*CUNY* | Supercompilation Compared with Partial Evaluation |
| **Gary L. Miller**<br>*Carnegie Mellon University* | Tree Contraction Parallel Algorithms |
| **Larry Rudolph**<br>*IBM Watson Research Ctr.* | Search for the Right Model |
| **Zvi Kedem**<br>*New York University* | Parallel Program Transformations for Resilient Computation |
| **Tom Cheatham**<br>*Harvard University* | Using the E-L Kernel Based Compiler System |
| **David Lillie**<br>*Imperial College* | Synthesis of a Parallel Quicksort |
| **Helmut Partsch**<br>*Catholic University of Nijmegen* | Experiments in Transforming Towards Parallel Execution |
| **Krishna Palem**<br>*IBM TJ Watson* | Algorithms for Instruction Scheduling on RISC Machines |
| **Apostolos Gerasoulis**<br>*Rutgers University* | A Fast Static Scheduling Algorithm |
| **Rob Strom**<br>*IBM Watson Research* | Optimistic Program Transformations |
| **Martin S. Feather**<br>*USC/ISI* | Constraints, Cooperation, and Communication |
| **Fritz Henglein**<br>*DIKU* | A Parallel Semantics for SETL |
| **Carolyn Talcott**<br>*Stanford University* | An Operational Approach to Program Equivalence |
| **David S. Wile**<br>*Information Sciences Inst.* | Integrating Syntaxes and Their Associated Semantics |
| **Allen Goldberg**<br>*Kestrel Institute* | Refining Parallel Algorithms in Proteus |

# SYSTEM DEMONSTRATIONS

| | |
|---|---|
| **Doug Smith**<br>*Kestrel Institute* | KIDS |
| **Allan Yang**<br>*Yale University* | Crystal and Metacrystal |
| **Neil Jones**<br>**Fritz Henglein**<br>*DIKU* | SIMILIX |
| **Valentin Turchin**<br>*CUNY* | The Refal Supercompiler: Demonstration of Function |
| **Jiazhen Cai**<br>**Bob Paige**<br>*New York University* | RAPTS |

# Preface

This book contains selected papers from the ONR Workshop on Parallel Algorithm Design and Program Transformation that took place at New York University, Courant Institute, from Aug. 30 to Sept. 1, 1991. The aim of the workshop was to bring together computer scientists in transformational programming and parallel algorithm design in order to encourage a sharing of ideas that might benefit both communities. It was hoped that exposure to algorithm design methods developed within the algorithm community would stimulate progress in software development for parallel architectures within the transformational community. It was also hoped that exposure to syntax directed methods and pragmatic programming concerns developed within the transformational community would encourage more realistic theoretical models of parallel architectures and more systematic and algebraic approaches to parallel algorithm design within the algorithm community.

The workshop Organizers were Robert Paige, John Reif, and Ralph Wachter. The workshop was sponsored by the Office of Naval Research under grant number N00014-90-J-1421. There were 44 attendees, 28 presentations, and 5 system demonstrations. All attendees were invited to submit a paper for publication in the book. Each submitted paper was refereed by participants from the Workshop. The final decision on publication was made by the editors.

There were several motivations for holding the workshop and for publishing papers contributed by its participants. Transformational programming and parallel computation are two emerging fields that may ultimately depend on each other for success. Perhaps, because ad hoc programming on sequential machines is so straightforward, sequential programming methodology has had little impact outside the academic community, and transformational methodology has had little impact at all. However, because ad hoc programming for parallel machines is so hard, and because progress in software construction has lagged behind architectural advances for such machines, there is a much greater need to develop parallel programming and transformational methodologies. This book seeks to stimulate investigation of formal ways to overcome problems of parallel computation - with respect to both software development and algorithm design. It represents perspectives from two different communities - transformational programming and parallel algorithm design - to discuss programming, transformational, and compiler methodologies for parallel architectures, and algorithmic paradigms, techniques, and tools for parallel machine models.

Computer Science is a young field with many distinct areas. Some of these areas overlap in their aims, differ distinctly in their approaches, and only rarely have constituents in common. Throughout the workshop the two (mostly nonoverlapping) communities in algorithms and in transformations reached for understanding, but also argued tenaciously in a way that reflected misunderstanding. It is not clear whether a bridge was formed between algorithms people and their counterparts in programming science. But the editors are optimistic. The chapters of this book sometimes show evidence of entrenchment, but they also reveal a synthesis of thinking from two different perspectives.

Certainly, there must be differences in the activities of researchers in algorithm design and program development, because these areas have different goals. In some respects the chapters of Rajasekaran and Reif, and also Langston are prototypical algorithms papers. Their goal is to compute a mathematical function. Their approach is to form this function by composition, parameter substitution, and iteration from other functions that are either already known to be computable within some time bound, or are shown in the paper to be computable within some bound. The functions being manipulated by algorithm designers are often thought not to be conveniently rendered with formal notation. For example, there may not be a single algorithm paper that makes explicit use of a higher order function (as we see in Pepper's chapter), and yet algorithms are invented every day in the algorithms community that are more complicated than any program conceived by researchers in functional programming. Because the algorithms community does not normally feel responsible for implementations, formal notation (which is so important when communicating with a machine) can be dispensed with. One cannot expect this approach to yield a reasonable program easily, but it may stimulate interest in producing such a program.

In the first four chapters, the transformational programming community is represented. The aim of this community is to make contributions to the science of programming. Particular concerns are with how to specify perspicuous mathematical programs, how to map these programs by meaning-preserving source transformations into efficient implementations, how to analyze the resource utilization of these implementations, and how to prove the specification, the transformations, and the implementation correct. Formal reasoning using notation systems is used so that programs can be manipulated systematically using calculi, and especially calculi that admit to computer mechanization. The approach in the transformational programming community is also genetic or top-down in the sense that the design, analysis, and correctness proof are integrated and developed together (as opposed to the classical verification approach). The stress on a correct implementation makes formalism and notation necessary - machines require precision and don't usually self-correct.

There are also common themes expressed among these chapters. Certainly, each contributor shows concern for practical results, and for theories rich in applications. The transformational chapters describe common goals for further mechanization of the transformational methodology, for transformations that capture algorithm design principles, and for transformational systems to be powerful and convenient enough to facilitate the simultaneous design of new algorithms and the development of their implementations. Within the two algorithm chapters we see an interest in notations for specifying algorithm schema and for the use of standard set theoretic notations for communicating problem specifications. Both communities strive for improvement at the meta-level. The stress on algorithm design principles within the algorithm community corresponds closely to the emphasis on transformational methodology within the transformational community. Consequently, the best solution is not the one most specialized to the particular problem, but one that is general enough to provide new meta-level thinking that might help solve other related problems.

Peter Pepper's provocative opening chapter argues sharply in favor of a transformational perspective for building provably correct parallel programs "by construction" on a realistic parallel architecture - namely the SFMD (single function instead of single instruction, multiple data, distributed memory). Based on practical considerations, he assumes that the data vastly exceeds the number of processors. Consequently, the problem of data distribution is the crucial part of program development. His 'deductive' approach is eclectic in the sense that some backwards reasoning in the style of formal verification is admitted. This approach begins with a high level functional sequential specification that is transformed into an equivalent high level parallel form. The use of infinite stream processing and other generic transformations (justified by associative and distributive laws) are used to implement the parallel specification efficiently. The method is illustrated with well selected examples (such as the fundamental prefix sum problem, and the more elusive problem of context free language recognition) that have also stimulated the algorithms community.

In Pepper's approach, the meta-level reasoning to produce datatype theories that support transformations and the reasoning behind the selection of these transformations are placed within a programming methodology (i.e., is part of a manual process) that utilizes the Bird-Meertens formalism. Pepper envisions an interactive system that supports the deductive approach in order to produce a specification at a low enough level of abstraction to be compiled by an automatic system (such as the one proposed by Gerasoulis and Yang) into efficient nets of communicating processes.

Doug Smith's chapter, which is methodologically similar to Pepper's, makes Pepper's proposal more credible by illustrating formal transformational derivations using KIDS, a working transformational programming system. Taking the derivation of Batcher's even-odd sort as an extended case study, Smith provides compelling examples of meta-level reasoning in the formation and manipulation of theories used to derive transformations. These theories are both specific to the domain of sorting, and generic in the way they capture the algorithmic principle of divide and conquer. Although meta-level reasoning is only partially implemented in KIDS, it is intriguing to consider the possibilities of a transformational system, which like LCF, has theories as objects.

Helmut Partsch addresses another important pragmatic concern - the reuse and portability of derivations to different parallel architectures. Like the previous two authors Partsch starts out with a functional specification. However, Partsch's method seeks to transform this specification into a lower level functional form defined in terms of a parameterized abstract parallel machine whose primitives are higher order library functions (called skeletons by Darlington et al. at Imperial College) that can be mechanically turned into implementations on a variety of specific parallel architectures. Partsch illustrates his methodology with a derivation of a parallel implementation of the Cocke, Kasami, Younger nodal span parser, which is precisely the kind of problem likely to catch the attention of the algorithms community. Parallel parsing is one of the big open algorithmic problems, and the so-called nodal span methods seem to offer greater opportunities for parallelism than other more popular sequential methods.

The chapter of Alberto Pettorossi, Enrico Pietropoli and Maurizio Proietti combines the interests of both the transformational and algorithm community by using program and transformation schemata and formal complexity analysis to prove optimality of their transformations. They investigate the difficult problem of directly implementing program schema containing first order nonlinear recursion on a synchronous parallel machine. The idea is to avoid (or at least bound) redundant computation of function calls and of subexpressions in general. The so-called tupling strategy (a generalization of the well known method of pairing in logic) of Burstall, Darlington, and Pettorossi is one of the main techniques. This results in what Gerasoulis and Yang call coarse-grain parallelism. Their Theorem 2 for general recursive programs proves that their rewriting scheme supports greater parallelism than the one described in Manna's book of 1974. The technical theorems in section 5 prove the correctness of an optimal parallel translation (the Eureka procedure) for a more particular recursive program scheme (which still represents a large class of functions). This scheme explains the earlier work of Norman Cohen in a more general setting, and also yields more efficient parallel implementations than does Cohen's transformations.

The transformational community is concerned with mechanical support for a largely intuitive and manual process of designing and applying transformations to obtain the highest level of program specifications that can be usefully compiled automatically into an executable form. The corresponding goal of the compiler community is to elevate the abstract level at which programs can be effectively translated into efficient executable forms. Thus, the results of the compiler community improve the fully automatic back-end of the full program development process envisioned by transformational researchers.

The chapter by Gerasoulis and Yang, which shows how to statically schedule (i.e. compile) partially ordered tasks in parallel, fits well into the current book. These authors are concerned with the problems of recognizing parallelism, partitioning the data and the program among processors, and scheduling and coordinating tasks within a distributed memory architecture with an asynchronous message passing paradigm for communication. Like so many important scheduling problems, these problems are NP-hard, so heuristic solutions are needed. The heuristics, which take communication complexity into account, favor coarse grain parallelism. The scheduling methods are illustrated with the problem of Gaussian Elimination. The authors give experimental results obtained using their PYRROS system.

The final two chapters are contributions from the algorithm design community. Both papers are unusual in being motivated by pragmatic concerns. Rajasekaran and Reif illustrate how randomization can make algorithms simpler (and hence simpler to implement) and faster using parallel algorithms derived for selection and sorting. They make use of notation to describe parameterized program schemas, where the choice of parameters is based on solutions to recurrences analyzing divide and conquer strategies. The notation is informative and enhances readability, but is not essential to the paper. The authors' methods are generally applicable to various parallel architectures, and not limited to PRAM models.

Langston surveys his own recent algorithms for merging, sorting, and selection in optimal SPACE as well as time per processor. These basic solutions are then used to provide optimal space/time solutions to set operations. The weaker but also more realistic PRAM model known as EREW (exclusive read, exclusive write) is used. Langston considers communication complexity, where a constant amount of space per processor implies optimal communication complexity. Langston also reports how an implementation of his merging algorithm on a Sequent Symmetry machine with 6 processors is faster than a sequential implementation even with small input instances.

During the workshop Donald Johnson asked whether transformational calculi can make the teaching of algorithms easier. Can language abstraction and formal transformational derivation help explain difficult algorithms?

Rajasekaran and Reif show how even a modest use of notation can improve readability considerably. In responding to Johnson's question at the workshop, Doug Smith thought that the overhead required to teach transformational methodology might be too costly for use in a traditional algorithm course. Chee Yap questioned whether pictures might be more useful than notation and algebraic manipulation for explaining some complicated algorithms in computational geometry. Johnson also raised the big question in many of our minds whether a transformational methodology or its implementation in a system such as KIDS could help in the discovery of an as-yet unknown algorithm? Smith reported that new algorithms have been discovered using KIDS, although KIDS is not primarily used for that purpose. Pepper seems to confront this issue with his parsing example. Pettorossi uses the compiler paradigm to obtain new results with program schemata in another attempt. However, these questions remain open.

The primary audience for this book includes graduate students and researchers in parallel programming and transformational methodology. Each chapter contains a few initial sections in the style of a first year graduate text-book with lots of illustrative examples. It could be used for a graduate seminar course or as a reference book for a course in (1) Software Engineering, (2) Parallel Programming, or (3) Formal Methods in Program Development.

Bob Paige, *New York University*
John Reif, *Duke University*
Ralph Wachter, *Office of Naval Research*

# 1

# Deductive Derivation of Parallel Programs

Peter Pepper

Fachbereich Informatik, Technische Universität Berlin, Germany

**Abstract**

The idea of a rigorous and formal methodology for program development has been successfully applied to sequential programs. In this paper we explore possibilities for adapting this methodology to the derivation of parallel programs. Particular emphasis is given to two questions: How can the partitioning of the data space be expressed on a high and abstract level? Which kinds of functional forms do lead in a natural way to parallel implementations?

## Prologue

*In the "Labour Day Workshop" – the proceedings of which are the contents of this book – two subcultures of computer science were brought together: "complexity theorists" and "formal algorithmists". Whether such a culture clash has any impact depends highly on the willingness to listen openly to the other side's viewpoints (the most elementary indication of which is, for example, the politeness to stay in the room when "the others" are talking). As far as I am concerned, the workshop was very valuable: On the one hand, it confirmed some of my earlier convictions, substantiating them with further evidence, but on the other hand, it also gave me some new insights into a field, where I felt not so much at home previously. Some of these learning effects are reflected in the following sections.*

### Why Formal Methods Are Substantial

*When scanning through the vast literature on algorithms and complexity, one cannot help but be impressed by the ingenuity of many designs. And this is true for the*

*sequential as well as for the parallel case. But at the same time it is striking, how little thought is given to the correctness of the suggested programs.*

*The typical situation found in the literature (of which our references are but a tiny sample) is as follows: A problem is stated. A solution is sketched – often based on an ingenious insight – and illustrated by examples, figures, and the like. At some point, suddenly a program text is "falling from heaven", filled with intricate index calculations and an elaborate interplay of parallel and sequential fragments. And it is claimed that this program text is a realization of the just presented ideas – even though no evidence is given for the validity of this claim. Instead, the core of the presentation, then, is a detailed complexity calculation. (Not surprisingly, the complexity arguments are usually based on the informal description of the idea, and not on the real program text.)*

*One problem with this procedure is that any variation of the problem needs a completely new – and extensive – programming effort in order to bridge the gap between the vague sketch of ideas and the detail-laden code. Even worse, however, is the total uncertainty, whether the program actually implements the intended solution. (After all – as Alberto Pettorossi put it during the workshop – if you don't care about the correctness of the result, any problem can be "solved" by a constant-time program.) Put into other words: Having the right answer a little later is better than having the wrong answer early.*

## ... And Why Complexity T h e o r y Is Only Marginal

*Knowing the actual costs of an algortihm is a decisive quality criterium, when assessing the results of different program derivations. And it is even reassuring to know that one is not too far away from the theoretical optimum. (Even though in practical engineering it is often enough to know that you are better than the competition.)*

*However, theorems of the kind "parsing can be done in $O(\log^2 n)$ time using $n^6$ processors" – which was seriously suggested as an important result during the workshop – barely can be considered relevant in an area, where n easily means several thousand symbols. But even when no such unrealisitc numbers occur, the "big-O-notation" often hides constants that actually render a computation a hundred times slower than was being advertised.*

## ... And What We Conclude From These Observations

*We develop programs – be they sequential or parallel – with two distinct goals in mind:*

- *Our foremost concern is the* c o r r e c t n e s s *of the program. This is the fundamental requirement for the technical realization of all our derivation steps. In other words, the correctness concern determines* h o w *we do things.*

- *But we also strive for the* e f f i c i e n c y *of our solutions. This is the motivation and strategic guideline for our derivations. In other words, the efficiency concern determines,* w h y *we do things.*

*Moreover, in software engineering there are also other concerns to be taken into account. Suppose, for instance, that there is one solution, which is well readable and understandable, but does not quite reach the optimum in performance, and that there is a competing solution, which does achieve the optimum, but is hard to comprehend. Then the preference will mostly be given to the first candidate, because the questions of modifiability and reusability are usually predominant in practice. Therefore, we strive for comprehensible developments of programs that exhibit a reasonably good perfromance; but we do not fight for the ultimate in efficiency. The means for performing such developments are the topic of this paper.*

Note: *To preclude a common misunderstanding from the very beginning: We do not want to invent new, as yet unknown algorithms; indeed, we do not even care, whether we apply a case study to the best, second best, or fifth best among the existing algorithms. All we are interested in is the derivation method.*

# 1 About the Approach Taken Here

We want to demonstrate here that the methods that are used in the formal derivation of sequential programs are equally applicable in the derivation of parallel programs. And the decisive effect is – in both cases – that the outcome of the process is *correct by construction.*

Of course, the individual rules that are used in the derivation of parallel programs differ from those used for sequential programs. And it is one of the aims of this paper to find out, what kinds of rules these are. The way, in which we strive for this goal, has proved worthwhile in the extensive research on sequential-program development: We take known algorithms and try to formalize the reasoning that went into their design.

Much has been written about the importance of producing *formally verified* software. And it is also known that the methods and tools for achieving this objective still are research topics. The various techniques suggested in the literature can be split into two main categories:

- *Verification-oriented techniques.* There is a specification, and there is a program. What is needed is a proof that the latter conforms to the former. The problem with this concept clearly is: Given a specification, how do we obtain a program, for which the subsequent proof attempt will not fail? And the answer is: Develop program *and* proof hand-in-hand. (An excellent elaboration of this idea is presented e.g. in the textbook of Gries [15].)
- *Transformation-oriented techniques.* There is a specification. To this specification a series of transformations is applied, aiming at a form that is executable on a given machine. The individual transitions in this process are performed according to formal, pre-verified rules; thus, the program is

"correct by construction". (This idea is elaborated e.g. in the textbooks of Bauer and Wössner [4] or of Partsch [19].)

As usual, the best idea probably is to take the valuable features from both approaches and combine them into a mixed paradigm. What we obtain in this way is a method, where transformation steps interact with verification efforts. But we still emphasize two central aspects: the *stepwise derivation* of the program from a given specification, and the *formally guaranteed correctness* of the derived programs. Therefore we baptize this method *"deduction-oriented"*.

Yet, there still remains the issue of *presenting* a program derivation to the reader (e.g. to a student, who wants to learn about algorithms, or to a customer, who wants to understand the product that he is about to purchase, or to a programmer, who has to modify the software). This presentation must be a compromise between formal rigour and understandability. Therefore we try to obey the following guidelines here (which are a weakened form of the idea of "literate program derivation" studied by Pepper [23]):

*The presentation of a program development should follow the same principles as the presentation of a good mathematical proof or of a good engineering design. We provide "**milestones**" that should be chosen such that*

- the overall derivation process can be understood by both the author and the reader, and
- the detailed calculations connecting any two successive milestones are fairly obvious and can be filled in (easily), maybe even by an automated transformation or verification system.

We will not indulge into a theoretical elaboration of the underlying principles of *deductive programming* here, but will rather present a collection of little case studies, which together should convey the overall idea. The technical details will be introduced along with these examples.

In the following subsections we briefly sketch some of the principles on which we base our case studies.

## 1.1 Concerning Target Architectures

The diversity of architectures makes parallel programming so much more confusing than sequential programming with its unique von Neumann target. Therefore it is unrealistic to believe that a single design can be ported automatically and efficiently across architectures. This is one of the benefits of our method: Since it is not product-oriented but rather derivation-oriented, we need not adapt final programs but may rather adapt developments, thus directing them to different architectures.

Nevertheless, we have to pick a reference model here in the interest of concreteness. For this reference model we make the following choices:

- MIMD architecture (*multiple-instruction, multiple-data*);
- distributed memory;

- $p \ll N$, where $p$ is the number of available processors, and $N$ is the size of the problem.

Since these assumptions are again in sharp contrast to the usual PRAM-setting (*parallel random-access memory*) chosen in complexity-oriented approaches, some comments seem in order here:

It is generally agreed that the PRAM model renders programming much simpler. Unfortunately, the assumption of a cost-free, homogeneous access of thousands of processors to one single shared memory is not realized by any existing or foreseeable machine. Some newer results that PRAM can be emulated in distributed memory through "randomization" are, unfortunately, again countered by practitioneers with the observation that random data distribution is about the worst that can happen in real machines. (This is a typical instance of the situation that a clever mathematical trick for saving the "big-O-category" introduces big constants into the "O", thus making the result practically irrelevant.) Some of these inherent problems can be found in several of the papers edited by Bode [6].

From this we conclude that there are no miraculous escapes from the hard tasks: *We have to deal with data distribution explicitly.* And a considerable part of our efforts will be devoted to this problem.

As to the number of processors, it is evident that for virtually all problems that are worth the investment of a parallel machine, the problem size $N$ will go into the millions, thus by far exceeding any conceivable machine size. Hence, it is only out of some curiosity that one may ask for the optimal number of processors. The practically relevant question always is, how the problem can be distributed over an arbitrary, but fixed number $p$ of processors. Nevertheless, the simpler situations of $p \approx N$ or even $p > N$ are included as borderline cases in our derivations.

## 1.2 Concerning Programming Concepts

The long experience with formal program development has shown that *functional programming concepts* are best suited for this methodology. Therefore we will follow the same route here, too, and perform all our deductions within a functional style.

In the sequential case the derivations are usually targeted towards specific recursion forms, known under the buzzword "tail recursion", possibly "modulo associativity", "modulo constructors", or "modulo inversion"; for details we refer to the textbooks by Bauer and Wössner [4] or Partsch [19]. On this level, compilers can take over, as is demonstrated by languages like HOPE, ML, or OPAL (see e.g. Pepper and Schulte [21], or Schulte and Grieskamp [26]).

In the parallel case we need some additional concepts. Not surprisingly it turns out that the construct (where $f$ is a function and $S$ is a data structure)

`Apply f ToAll S,` denoted here in the form `f✯S`

plays a central role in this context. As we will see later on, the individual components of the data structure $S$ correspond to the *local memories* of the individual processors. And in many cases the function *f* actually is of the form $f \equiv g(x)$, where *g* is some higher-order function. That is, we have situations of the kind

`Apply g(x) ToAll S,` denoted here in the form `(g x)☆S.`

Then *x* corresponds to the data that has to be *communicated* among processors. These two remarks should already indicate that we can indeed perform the essential development steps on the functional level. To work out the details of this concept is a major goal of this paper.

We mainly use here the following operators[1]:

```
f☆S            -- apply-to-all
⊕☆(S1,S2)      -- zip (apply-to-all for binary functions)
⊕/S            -- reduce
```

*Explanation*: We characterize the above operators for sequences; but they work for many other data structures analogously.

- The ***apply-to-all operator*** '`f☆S`' applies the function `f` to all elements in the sequence `S`. For instance:

  `double☆ <3,4,5>  = <6,8,10>.`

- The ***zip operator*** applies its operation pairwise to all the elements of its two argument sequences. Due to its close analogy to the apply-to-all operator we use the same symbol '☆' here. For instance:

  `+☆ (<1,5>, <2,4>) = <3,9>.`

- The ***reduce operator*** '`/`' connects all elements in the sequence `S` by means of the given operation. For instance:

  `+ / <3,7,9> = 3+7+9.`

  For the empty sequence we define `+/<> = 0`. More generally, we use the neutral element of the corresponding operator (the existence of which is ensured in all our examples).

In order to save extensive case distinctions, it is sometimes very convenient to extend the zip-operator to sequences of different lengths:

*Convention*:

When the zip-operator is applied to sequences of different lengths, then it acts as the identity on the remainder of the longer sequence. For instance:

`+☆ (<1,5>, <3,7,2,9>) = <4,12,2,9>.`

In order to describe also the communication aspect explicitly, we use so-called *stream-processing functions*: A ***stream*** is a potentially infinite sequence of elements with a non-strict append function (see e.g. Broy [7]). Note, however, that this is not mandatory for our approach. Most other styles of expressing parallel programs would do as well. However, from our experience with sequential-program

---

1 This style of working has become known in certain subcultures under the buzzword "Bird-Meertens style" (see Bird [1989] or Lambert Meertens' contribution in this volume); to others it may be reminiscent of certain features of APL.

development we know that functional languages are much better suited from the point of view of elegance.

We refrain from listing our other programming concepts here, but rather introduce them as needed during the presentation of our subsequent case studies. (Darlington et al. [10] give a more extensive list of useful operators.)

## 1.3 Concerning Development Strategies

The orientation towards *MIMD architectures with distributed memory* requires that we strive for a large-grain parallelism. This is always reflected in an appropriate *partitioning of the data space*. And this partitioning may take place either for the argument data of our functions or for their result data.

On the other hand, it is evident that massive parallelism can only come from applying the same function almost identically to huge numbers of data items. (It is obviously not feasible to write software that consists of thousands of different but simultaneously executable functions.) At first sight, this seems to call for an SIMD paradigm. But at closer inspection it merely shows that the MIMD/SIMD distinction is too weak for methodological considerations. What we do need is a concept that could be termed

- *"SFMD"-paradigm* (*single-function, multiple data*).

Here, "single-function" means that a (potentially very large) function is executed on many processors, but on every processor with another set of data, and in a more or less unsynchronized manner. (This is sometimes also called "single-program, multiple-data" paradigm.)

Technically, our approach usually proceeds through the following steps:

- We start from a formal *problem specification*.
- Then we derive a (sequential) *recursive solution*.
- The decisive step then is the choice of a *data space partitioning* (which is usually motivated by concepts found in the sequential solution).
- Then we adapt the solution to the partitioned data, leading to a *distributed solution*.
- Finally, we implement this distributed solution by a *network of parallel processes*.

We should point out that this way of proceeding differs in an important aspect from traditional approaches found in literature. There, one often starts from a – hopefully correctly derived – sequential solution and tries to parallelize it (maybe even using special compilers). By contrast, we transform the high-level functional solution and then implement the thus derived new version directly on parallel machines.

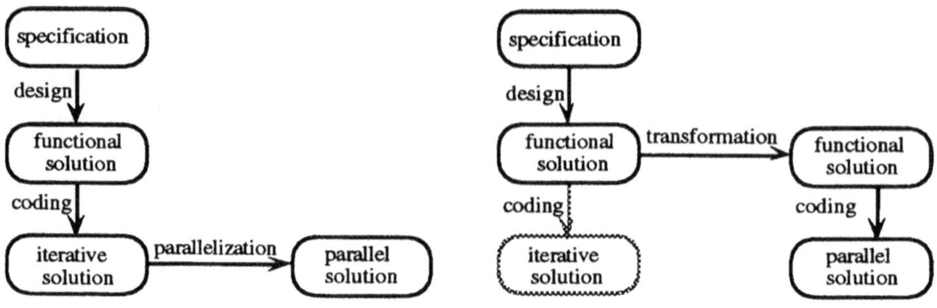


traditional approach approach taken here

# 2 An Introduction And An Example: Prefix Sums

The computational structure of the following example can be found in various algorithms, such as evaluation of polynomials, Horner-schemas, carry-lookahead circuits, packing problems, etc. (It has even been claimed that it is the most often used kind of subroutine in parallel programs.) For our purposes the simplest instance of this class will do, viz. the computation of the "prefix sums". Due to its fundamental nature, treatments of this example can be found at various places in the literature (including textbooks such as those of Gibbons and Rytter [13] or Cormen et al. [9]). Note, however, that most of these treatments produce another algortihm than ours. We will come back to this comparison in section 8.

*This is the one case study that we present in considerable detail. The other examples will be more sketchy, since they follow the same principles. In the following derivation, the milestones are given in subsections 2.1 and 2.2, and some representative examples of the detailed calculations are given in 2.3.*

## 2.1 Sequential Solutions

In order to get a better understanding of the problem we first develop some sequential solutions. Then we adapt these ideas to the parallel case.

***Milestone 1:*** *Specification.* We want to solve the following problem: Given a nonempty sequence of numbers `A`, compute the sequence of numbers `Z` such that `(Z i) = Sum(A 1..i)`. For instance,

```
(psums <1,2,3,4,5>) = <1,3,6,10,15>
```

This is formalized in the specification

```
1. Initial specification:
-----------------------------------------------------------
FUN psums: seq[num] → seq[num]

SPC (psums A) = B   ⇒    (B i) = (sum (A 1..i))
                         for i = 1..#A
```

Here `(B i)` denotes the i-th element of the sequence `B`, `(A i..j)` denotes the subsequence ranging from the i-th to the j-th element, and `#A` denotes the length of `A`. The keyword `FUN` designates a functionality, and the keyword `SPC` a specification. For function applications we often employ a Curried notation; that is, we write `(f x y)` instead of `f(x,y)`; parantheses are used where necessary for avoiding ambiguities or enhancing readability. The function application `(sum S)` yields the sum of all the numbers in the sequence `S`, that is,

```
(sum S) = +/S.
```

***Milestone 2:*** *A recursive solution.* Our first solution is straightforward:

```
2. A recursive solution
-----------------------------------------------------------
DEF (psums empty) = empty

DEF (psums A^a)   = (psums A)^(sum A^a)
```

The keyword `DEF` signals an executable definition, which may be given in several parts, using "call-by-pattern". The operation '^' appends an element to a sequence. As a convention, we always use capital letters for sequences and small letters for elements. This is just a visual aid for the reader; an ML-like type inference can deduce this information from our programs without falling back upon such conventions. Note that the subterm `(sum A^a)` could be replaced by the equivalent one `(sum A)+a`, since '+' is associative.

***Milestone 3:*** *Left-to-right evaluation of sequences.* We usually read sequences from left to right; therefore it may be helpful to consider a variant of our function, which takes this view.

```
3. An alternative solution
-----------------------------------------------------------
DEF (psums empty) = empty

DEF (psums a^A)   = a^((a+)☆(psums A))
```

In this solution we have lost the expensive calculation of `(sum A)` in each recursive incarnation, but at the price of introducing an apply-to-all operation. Hence, the complexity is $O(n^2)$ in both solutions. Note that we use the symbol '^' also for prepending an element in front of a sequence; but this overloading does not do any harm.

***Milestone 4:*** *Improving the recursive solution.* The computation of `(sum A)` in variant 2 above or the computation of `(a+)`☆ in variant 3, respectively, can be avoided by maintaining a suitable carry element as an additional parameter. (The pertinent transformation is often called "strength reduction" or "finite differencing".) This brings us to the desired O($n$) algorithm.

```
4. Recursive solution with carry

FUN ps: num → seq[num] → seq[num]

DEF (ps c empty)  = empty
DEF (ps c a^A)    = (c+a)^(ps (c+a) A)

LAW (psums A) = (ps 0 A)
```

The keyword `LAW` denotes a valid fact about the given definitions. In our case, the property expresses an equivalence between the original and the derived function. (In a logical framework, the definitions therefore play the role of axioms, and the laws play the role of theorems.)

## 2.2 Parallel Solutions

*The above derivation is still oriented at a sequential programming model. Now we want to re-orient it towards a parallel programming model. But for the time being we still remain on an abstract functional level.*

*As we will show in section 3, the following derivation is nothing but an instance of a certain transformation rule. In other words, this whole subsection is actually superfluous, because we have general rules at our disposal that bring us directly from the recursive solutions above to corresponding parallel implementations. But we nevertheless present the following development steps, because they may help to motivate the abstract rules to be presented afterwards.*

*As usual, we content ourselves with the milestones here, deferring the formal calculations inbetween to a later section.*

In order to enable parallel computations we have to partition the data space appropriately. We decide here to split the sequence under consideration into blocks of size $q$. (As will turn out later, this size should be chosen such that $2q$-1 is the number of available processors.)

BB = 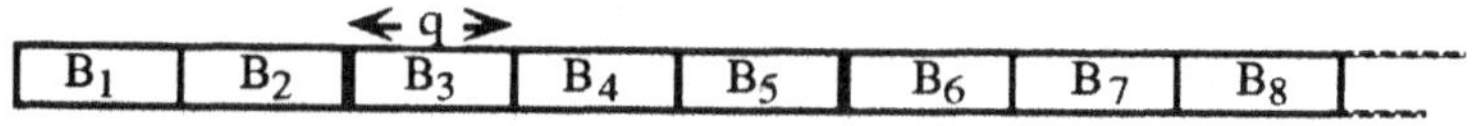


To simplify the presentation, we assume here that the size of the sequence `A` is just $k*q$; but this is not decisive for the algorithm.

***Milestone 5:*** *Partitioning of the data space.* As illustrated above, we pass from a sequence of numbers `A` to a sequence of sequences of numbers `BB`. The relationship between the two is, of course,

```
A = ++/BB
```

where '`/`' is the aforementioned reduce operator, and '`++`' denotes the concatenation of sequences. Hence, the *conc-reduce* '`++/`' just flattens a sequence of sequences into a single sequence.

| *Notational convention*: |
|---|
| `TYPE tuple = «seq with a fixed length»` |

That is, we speak of "tuples" if we want to emphasize that our sequences have a fixed (but arbitrary) length. Then we can formulate our program in the following form.

```
5. Recursive solution after partitioning

FUN PSUMS: seq[tuple[num]] → seq[num]

DEF (PSUMS empty)   = empty
DEF (PSUMS B^BB)  = (psums B) ++ ((sum B)+)☆(PSUMS BB)
```

*Explanation*: The effect here is that `((sum B)+)☆...` adds the sum of the elements of `B` to each element in `(PSUMS BB)`.

***Milestone 6:*** *Avoiding unnecessary additions.* If we look at the above solution from an operational point of view, we can immediately detect a source of unpleasant inefficiency: We add the sum of the first block $B_1$ to the whole remainder of the sequence `BB`; then we add the sum of block $B_2$ to the remaining sequence; and so on. Because addition is associative, we may instead collect these partial sums and apply them only once when generating the output blocks. This claims again for a carry element.

```
6. Partitioned solution with carry elements

FUN PS: num → seq[tuple[num]] → seq[num]

DEF (PS c empty) = empty
DEF (PS c B^BB) = ((c+)☆(psums B))++(PS (c+(sum B)) BB)
```

*If we look at the above solution, we can make the following observations: The recursive call of `PS` has an argument that depends on `B` – and this dependency precludes a simple parallel evaluation of the recursion. Hence, we have to look at the "local" computation in the body. Here the apply-to-all operator looks promising. Therefore, we will concentrate on this program part first.*

***Milestone 7:*** *Parallelization of the function body.* We want to compute the main subexpression of the body of PS; that is, we want to compute the sequence and the carry element

```
Z   = ((c+)☆(psums B))
c'  = c+(sum B) = (last Z)
```

Intuitively, it is immediately clear that the following network performs this calculation: The upper half yields the sequence (psums B), and the lower half adds the carry element c to each element of this sequence. (Note that the left upper addition is actually superfluous, but the uniformity of the design helps technically.)

7. Partitioned solution with carry elements

```
DEF (PS c empty)  = empty
DEF (PS c B^BB)   = Z ++ (PS c' BB)
                    WHERE Z,c' = Net(B,c)
```

Net(B,c):

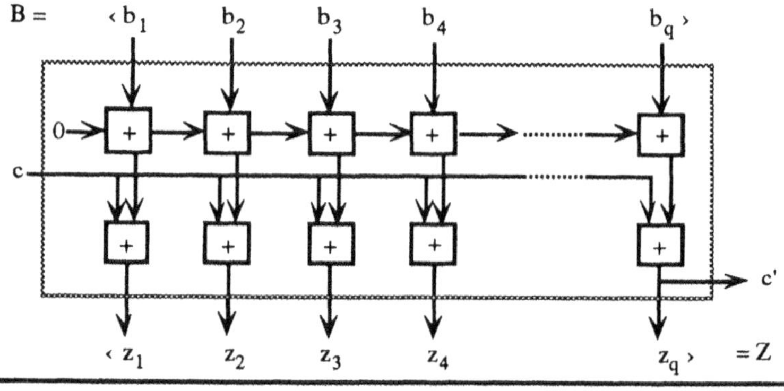


*Even though we have now arrived at a network solution for the body of our main function, a closer inspection reveals severe deficiencies. The network still contains a data flow from the very left to the very right. And this means that the parallelism is only superficially present; in reality the causal dependencies will effect a mostly sequential computation. This is the point, where we have to look across the boundaries of the recursion structure of our function* PS*. We will simply apply the following trick: Instead of letting the next incarnation of* PS *wait for the completion of the previous one, we let it start right away. The principles of stream processing then take care of potential race conditions and the like.*

***Milestone 8:*** *Full parallelization.* The underlying transformation principle is well known: Recursion equations for functions are turned into recursion equations for streams (see e.g. Broy [7]). Technically speaking, this entails the passing from a stream of tuples to a tuple of streams.

To help intuition, we sketch below the working of the network; the sequence BB is put into the network block by block, and the carry element flows around in a feedback loop. Note, however, that the $q$ input streams are not synchronous; due to the dependencies, the element $b_{q-1,1}$ enters the network together with the element $b_{1,q}$. For easier reference we have given the two kinds of addition nodes the different names P and S.

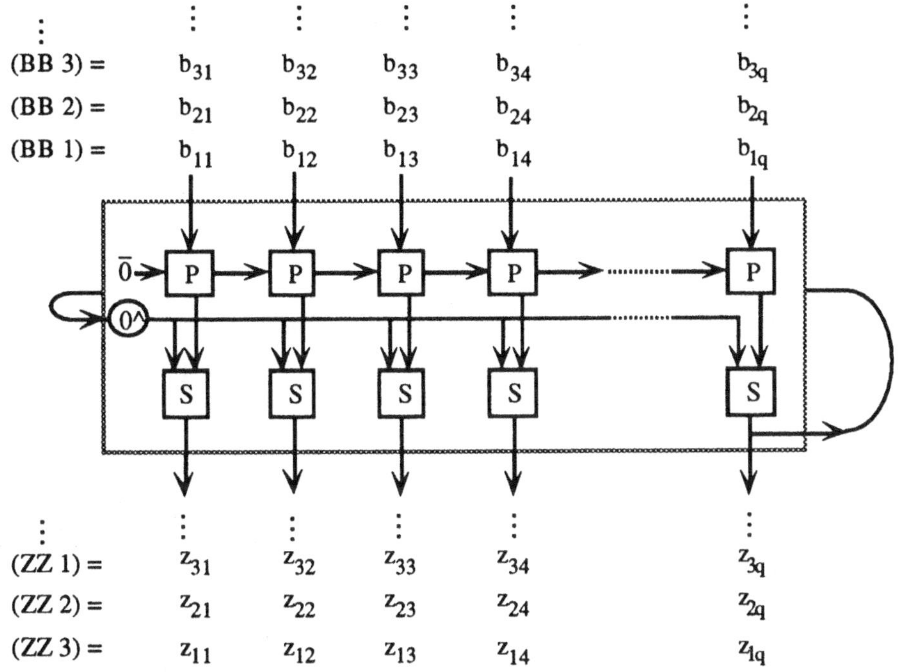


Here, the understanding of the node '0^' is that the element '0' is prepended in front of the incoming stream.

***Discussion:***

- *Obviously, we need 2q-1 processors for this solution (since $P_1$ is actually superfluous). The time complexity is in the order O(N/q+q).*
- *It is shown by Nicolau and Wang [17] that for a fixed number q<<N of processors this solution is close to the theoretically achievable optimum, the complexity of which is O(N/q+log q). However, the optimal solution is much harder to program (see section 8).*
- *This algorithm clearly exhibits the fine granularity of SIMD-parallelism.*
- *If we are in a distributed-memory MIMD-environment, then the initial sequence has to be properly distributed over the local memories – which may not be possible in all applications. (See also section 3.)*
- *Finally, we want to emphasize again that the complete derivation of this subsection is covered by a single transformation rule given in section 3 below.*

It remains to represent the above network in some kind of program notation. Below, we use an ad-hoc syntax, which should be self-explaining; it just describes the above graphical presentation of the network. Note that we now provide our functions with named parameters and results.

```
8. Network implementation of complete function PSUMS

FUN PSUMS: tuple[stream[num]] → tuple[stream[num]]

DEF PSUMS BB =
     LET q = size BB IN
    NET
     AGENTS (P 0) = «zeros»
            (P i) = P                       [i = 1..q]
            (S i) = S                       [i = 1..q]
     CONFIG (P i).top  = (BB i)             [i = 1..q]
            (P i).left = (P i-1).out        [i = 1..q]
            (S i).top  = (P i).out          [i = 1..q]
            (S i).left = 0^(S q).out        [i = 1..q]
     OUTPUT (S i).bot                       [i = 1..q]

-- auxiliary functions

FUN P: left:stream[num] × top:stream[num] →
                                       out:stream[num]
FUN S: left:stream[num] × top:stream[num] →
                                       out:stream[num]

DEF P(left,top).out = +☆(left,top)
DEF S(left,top).out = +☆(left,top)
```

*We do not indulge here any deeper into the issue of possible language constructs for the presentation of networks. But it seems plausible that we should aim for the same kind of abstraction that is found in the area of data structures. As a matter of fact, our networks can be treated fully analogously to data structures, with the additional complication that the communication patterns need to be described as well.*

## 2.3 Some Detailed Calculations

*In concluding this case study, we want to present some of the detailed calculations that are needed to fill the gaps between our various milestones. Since this shall merely be done for illustration purposes, we give these calculations only for some selected steps.*

*Doing this kind of calculations is like going back to the axioms of predicate calculus when proving some lemma of, say, linear algebra or analysis. As a consequence, the derivations tend to become tedious and quite mechanical. Therefore we may hope to leave them to automated systems some day.*[1] *Fortunately, we need not wait for the emergence of such systems, but still have the option of retreating to pencil and paper. But then a short notation is most welcome, and this is where the Bird-Meertens style comes in very handy. Derivations in this style usually employ certain basic properties of the underlying operators (which are elaborated in detail in the work of Bird and Meertens, as exemplified e.g. by Bird [5]); typical examples are (where '∘' denotes function composition, and '⊕' is an arbitrary binary operation)*

```
f★(g★A)    = (f∘g)★A                [a]
f★(a^B)    = (f a)^(f★B)            [b]
f★(A++B)   = (f★A)++(f★B)           [c]
f★empty    = empty                  [d]
⊕/(a^A)    = a⊕(⊕/A)                [e]
```

*The kinds of calculations that have to be performed in this style are reminiscent of algebraic calculations in, say, group theory: relatively technical and simple for the trained expert, but at the same time hard to follow for the uninitiated reader. Therefore we list at each stage the equation that has been applied. As to the strategies, we usually try to derive recursive instances of the function under consideration.*

*Derivation of milestone 2: A first recursive solution.* As a fundamental concept in the working with sequences we often have to employ the "sequence of initial sequences"; for instance,

```
(inits <1,2,3>) = <<1>,<1,2>,<1,2,3>>.
```

Definition of `inits`

```
FUN inits: seq[α] → seq[seq[α]]
DEF (inits empty) = empty                  [f]
DEF (inits A^a)   = (inits A)^(A^a)        [g]
```

For this function we have the following basic properties:

---

1 Work on such systems currently is a very active research area. A good example is the KIDS system (see the paper of D. Smith in this volume).

```
f★(inits A^a)  =  (f★(inits A))^(f A^a)              [1]
inits(A++B)  =  (inits A)++((A++)★(inits B))          [2]
```

Using the symbol Σ as an abbreviation for the function sum, that is

```
Σ A  =def  +/A,                                        [3]
```

we can rewrite the specification of psums as follows:

```
psums A  =def  Σ★(inits A)                             [4]
```

Now we are in a position to perform the following formal calculation that derives from the above definition a recursive equation for psums:

```
(psums A^a)  =  Σ★(inits A^a)                  [due to 4]
             =  Σ★((inits A)^(A^a))            [due to g]
             =  (Σ★(inits A))^(Σ A^a)          [due to 1]
             =  (psums A)^(Σ A^a)              [due to 4]
             =  (psums A)^((Σ A)+a)            [due to 3]
```

*Derivation of milestone 3: Left-to-right evaluation.* For the sake of completeness we also want to demonstrate the derivation of the second recursive solution. To this end, we have to employ the following basic property for the operation inits (where <a> stands for the singleton sequence consisiting only of the element a):

```
(inits a^A)  =  <a>^((a^)★(inits A))                   [5]
```

Then we can calculate another recursive equation for psums:

```
(psums a^A)  =  Σ★(inits a^A)                  [due to 4]
             =  Σ★(<a>^((a^)★(inits A)))       [due to 5]
             =  (Σ<a>)^(Σ★((a^)★(inits A)))    [due to b]
             =  a^(Σ★((a^)★(inits A)))         [due to 3]
             =  a^(Σ∘(a^))★(inits A))          [due to a]
             =  a^((a+)∘Σ)★(inits A))          [due to e]
             =  a^((a+)★(Σ★(inits A)))         [due to a]
             =  a^((a+)★(psums A))             [due to 4]
```

*Derivation of milestone 4: Introduction of a carry element*: Even though it is a standard application of so-called strength reduction, we want to demonstrate also this derivation. Our starting point is the following definition of the new function ps:

```
(ps c A)  =def  (c+)★(psums A)                         [6]
```

Then we obtain immediately the law that establishes the connection between psums and ps:

```
(ps 0 A)  =  (0+)★(psums A)  =  (psums A)
```

A recursive definition of ps itself is established by the following derivation:

```
(ps c a^A)
   = (c+)★(psums a^A)                           [due to 6]
   = (c+)★(a^((a+)★(psums A)))             [def. of psums]
   = (c+a)^((c+)★((a+)★(psums A)))              [due to b]
   = (c+a)^((c+)∘(a+))★(psums A)))              [due to a]
   = (c+a)^((c+a+)★(psums A))                [assoc. of +]
   = (c+a)^(ps (c+a) A)                         [due to 6]
```

*Derivation of milestone 5: The partitioned solution.* Now we want to calculate the partitioned solution, which is based on the sequence of sequences BB with the characteristic property

```
++/BB = A                                              [7]
```

The initial sequences of A – which are fundamental for our derivations – are then provided by the following function:

```
INITS BB ≝ inits A = inits(++/BB)                      [8]
```

For this function we claim that the following property holds:

```
INITS(B^BB) = (inits B)++((B++)★(INITS BB))            [9]
```

Proof of [9]:

```
INITS(B^BB)
   = inits(++/(B^BB))                           [due to 8]
   = inits(B++(++/BB))                          [due to d]
   = (inits B)++((B++)★(inits ++/BB))           [due to 2]
   = (inits B)++((B++)★(INITS BB))              [due to 8]
```

(End of proof)

Now we are ready to to derive the recursive version 4 of our algorithm PSUMS:

```
PSUMS(B^BB)
   ≝ psums(++/(B^BB))
   = Σ★(inits ++/(B^BB))                            [due to 4]
   = Σ★(INITS(B^BB))                                [due to 8]
   = Σ★((inits B)++((B++)★(INITS BB)))              [due to 9]
   = (Σ★(inits B))++(Σ★((B++)★(INITS BB)))          [due to b]
   = (psums B)++(Σ★((B++)★(INITS BB)))              [due to 4]
   =     ...      ++(Σ∘(B++))★(INITS BB))           [due to a]
   =     ...      ++((ΣB +)∘Σ)★(INITS BB))          [ arithm.]
   =     ...      ++((ΣB +)★(Σ★(INITS BB)))         [due to a]
   =     ...      ++((ΣB +)★(Σ★(inits(++/BB))))     [due to 8]
```

```
=      ...    ++((ΣB +)☆(psums(++/BB)))        [due to 4]
= (psums B)++((ΣB +)☆(PSUMS BB))       [by definition]
```

*These examples should suffice to illustrate, how the gaps between our milestones may be filled by detailed (and very technical) calculations.*

*Remark: Obviously, this notation is unsuited for writing large, well-documented software systems. But it is quite nice for doing detailed calculations in a compact fashion – provided that the milestones inbetween are sufficiently understandable. As a consequence, we actually want to work with at least two notations in our developments. This observation led to the concept of coexisting "local formalisms", as suggested by D. Wile [28] (see also his contribution to this workshop).*

# 3 Transformation Rules For Parallelization

*Now we abstract from the above example and consider the general program schemes behind it. Even though we generally work with finite sequences, let us mainly consider infinite streams for the moment (in order to shorten the presentation). The subsequent properties usually can be verified by an induction based on substream ordering (as elaborated by Broy [7]).*

Our goal is to identify certain standard patterns of high-level functional programs and to associate equivalent parallel algorithms to them. (Such patterns are sometimes referred to as "skeletons", for instance by Cole [8] or Darlington et al. [10].) In the sequel we derive rules for the following patterns (where the first three patterns are relatively standard in the literature, whereas the recurrences in the last two patterns appear to be less known):

- Pattern 1: `h☆`
- Pattern 2: `(h x)☆`
- Pattern 3: `(h/)☆(inits S)`
- Pattern 4: `DEF (F x^S) = (h x)☆(e^(F S))`
- Pattern 5: `DEF (F x^S) = (g x)^(F (h°g x)☆S)`

## 3.1 From the ☆-Operator to Streams and Nets

The simplest source for parallelization is the plain application of the apply-to-all operator '☆' to a given stream `S:stream[α]`. For the result stream `R:stream[α]` we obtain the following definition and corresponding network implementation:

```
DEF R = h☆S
```

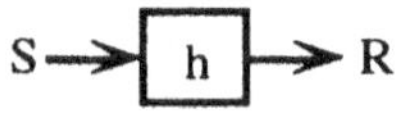


If `S` is finite, and if there are sufficiently many processors available, we can also unroll this network into the form (where $q$ = `#S`)

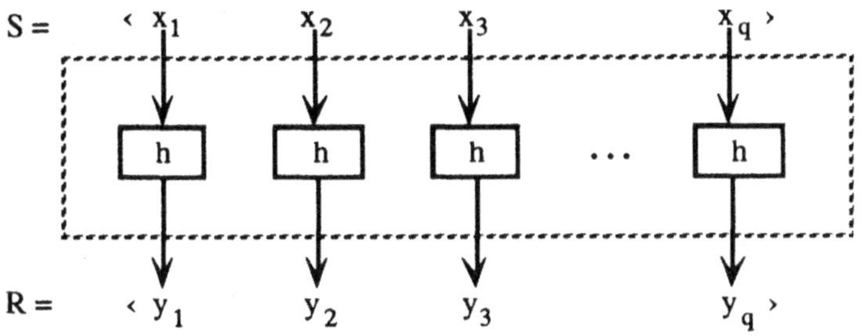


If there are not sufficiently many processors available, we have to assign to each processor a whole fragment of `S`. (This principle is often encountered under the buzzword of *Brent's theorem* in the literature.) Since there are at least two standard ways for breaking up the stream `S`, we first present an abstract setting using two operations `Split` and `Merge`, which are only partially specified for the time being.

```
FUN Split_q: stream[α]           → tuple_q[stream[α]]
FUN Merge:   tuple[stream[α]] → stream[α]
LAW Merge ∘ Split_q = Id
LAW Split_q ∘ Merge = Id
LAW h☆(Merge∘Split_q S) = Merge((h☆)☆(Split_q S))
```

Using these properties of `Split` and `Merge` we obtain the following transformation rules.

***Rule 1***

The apply-to-all operator

`h☆S`

is equivalent to the network

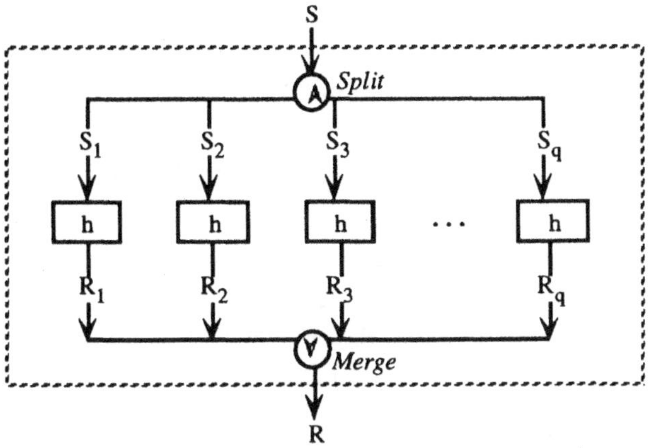

It is worthwhile to also consider the case, where the function h actually is a higher-order function depending on a data item x. Then we obtain a network, where the value x is *broadcasted* to all processors.

***Rule 2***

The apply-to-all operator with a higher-order function

(h x)☆S

is equivalent to the network (where $\bar{x}$ denotes the stream ‹x,x,x,...›)

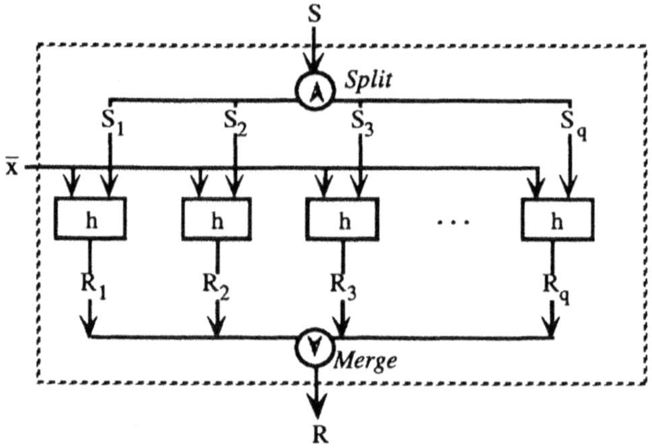


*These two rules allow us to evaluate an "apply-to-all" situation in parallel on a given number of q processors. Note that the length of S need not be a multiple of q, because the semantics of* `Split` *and* `Merge` *takes care of streams of different lengths.*

*We refrain from encoding the above network in textual form. The examples in section 2.2 should suffice as an illustration of how this could be done.*

For the design of suitable functions `Split` and `Merge` we have two obvious choices:

- If the stream S is finite, we can split it into $q$ blocks:

  $S = S_1{+}{+}S_2{+}{+}...{+}{+}S_q$

  These blocks usually will be of equal length, but this need not be the case. Then we have

  $(\text{Split}_q\ S) = \langle S_1, ..., S_q \rangle$

  $(\text{Merge}\ \langle S_1, ..., S_q \rangle) = S_1{+}{+}\ ...\ {+}{+}S_q$

- If the stream S is finite or infinite, we can partition it in such a way that

  $((\text{Split}_q\ S)\ i) = \langle (S\ i), (S\ i{+}q), (S\ i{+}2q), ... \rangle$

Then `Merge` performs an *interleaving*. More formally, we may describe this partitioning by the following two-stage process: First, we partition the stream `S` into blocks of size $q$ .

SoT =
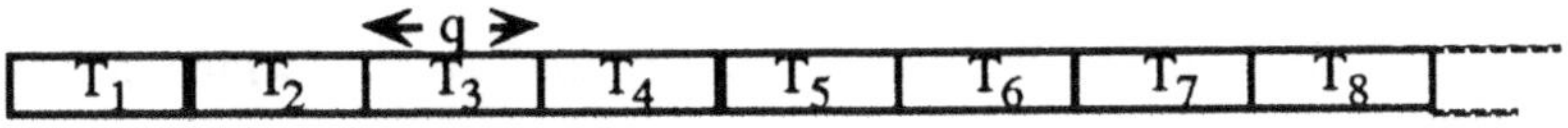


That is, we have a stream of tuples:

```
FUN SoT: stream[tuple[α]]
LAW ++/SoT = S
```

Now we can calculate the following relationship:

```
h✩S  =  h✩(++/SoT)  =  ++/((h✩)✩SoT).
```

The resulting stream of $q$-tuples is isomorphic to a $q$-tuple of streams. That is, we have the following transposition:

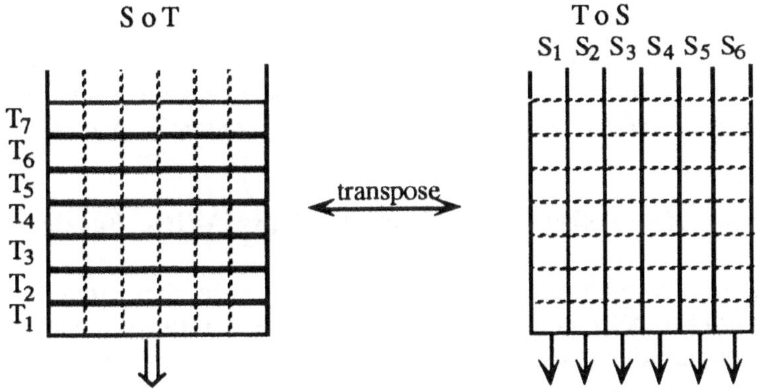


The relationship between these two data structures is specified as follows (where `transpose` actually stands for two overloaded functions)

```
FUN SoT: stream[tuple_q[α]]
FUN ToS: tuple_q[stream[α]]
LAW ToS = (transpose SoT)
            ⇔  ((SoT i) j) = ((ToS j) i)   for 1≤j≤q
LAW transpose∘transpose = Id
```

The relationship to the original stream `S` is therefore given by the following equality:

```
S = ++/SoT = ++/(transpose ToS)
```

On the basis of these definitions we obtain the following relationship:

```
h✩S  = h✩(Merge ∘ Split_q S)
     = h✩(Merge ToS)      WHERE ToS =def (Split_q S)
     = h✩(++/(transpose ToS))
     = ++/((h✩)✩(transpose ToS))
```

```
= ++/(transpose((h★)★ToS))
= Merge((h★)★ToS)
= Merge((h★)★(Split_q S))
```

This is the required property for `Split` and `Merge`, which makes our rules applicable.

*Discussion:*

*Whether these transformations are useful or not, depends on the context, in which they are used. The net only effects a true parallelization of work, when the operation h is much more costly than the atomic actions of distributing and recollecting elements within Split and Merge. This is for instance the case, when*

- *... the individual processors can fetch and store their stream elements directly from a global shared memory;*
- *... the individual streams are already predistributed over the local memories;*
- *... Split and Merge actually have to be implemented as sequential operations on streams, but h is a very expensive and time-consuming operation.*

## 3.2 From Recursive Functions to Nets With Feedback

The above transformations are very simple, because they are nothing but straightforward parallelizations of the apply-to-all operator. But, of course, there are also more complex situations, where recursive functions lead to nets with feedback loops.

*Introductory toy example*: As a preparation for the subsequent discussion consider the following trivial example:

```
FUN F: stream[α] → stream[nat]
DEF (F x^S) = succ★(0^(F S))
```

This function maps every stream `S` into the stream `<1,2,3,...>`. Now consider the recursive stream equation for `R:stream[nat]`

```
DEF R = succ★(0^R)
```

This is again the aforementioned stream `R=<1,2,3,...>`. Hence, we have the equation:

```
LAW R = (F S)        for any stream S.
```

The stream `R` is produced by the following net.

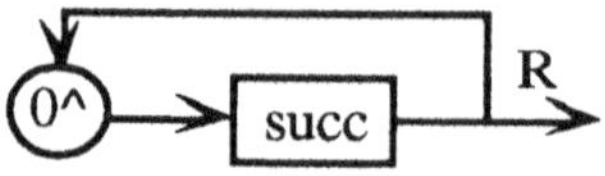


Such a network can be "unrolled" an arbitrary number of times (where the merge operator '≻' denotes again an exact interleaving of the input streams):

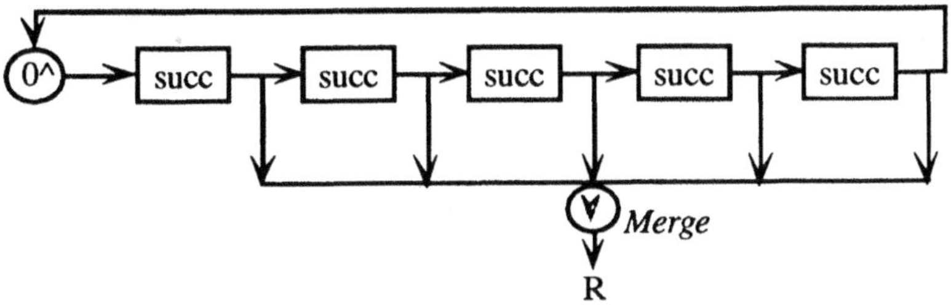


*Proof*: We generalize a little by using an arbitrary function h instead of succ and an arbitrary element c instead of 0. That is, we consider the stream equation

```
R = h☆(c^R)
```

Now let

$R_1, R_2$ = $\text{Split}_2$ R

Then a simple induction shows that the following relationships hold

$$R_1 = (h\ c)\hat{}(h☆R_2) \qquad [= (h\ c)\hat{}(h^2☆R_1)]$$
$$R_2 = h☆R_1 \qquad [= (h^2\ c)\hat{}(h^2☆R_2)]$$
$$R = \text{Merge}\ \langle R_1, R_2\rangle$$

The generalization from 2 to an arbitrary number *q* is straightforward, thus yielding the above net.

*(End of introductory example)*

Now let us consider a generalization which is more realistic. In the function psums we encounter the following pattern:

```
DEF (F S) = (h/)☆(inits S)
```

As we had seen in stage 3 of our psums derivation, this function leads to the recurrence relation

```
DEF (F x^S) = x^((h x)☆(F S))
```

If h has a *neutral element* e, that is, (h x e) = x, we can rewrite this definition into

```
DEF (F x^S) = (h x)☆(e^(F S))
```

This function produces the following stream:

```
LAW (F S) = R   WHERE   R = h☆(S,e^R)
```

which is generated by the network

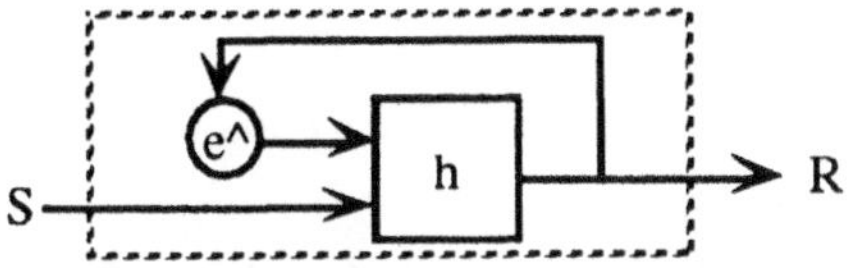


As before, this network can be unrolled. The proof from the above toy example only needs to be extended from $h☆R_i$ to $h☆(S_i, R_i)$ in order to yield the network

$$\text{DEF Net}(S_1,\dots,S_q) = R_1,\dots,R_q \quad \text{WHERE } R_1 = h\star(S_1, e\hat{}R_q)$$
$$R_2 = h\star(S_2, R_1)$$
$$\dots$$
$$R_q = h\star(S_q, R_{q-1})$$

In graphical form this network may be represented as follows:

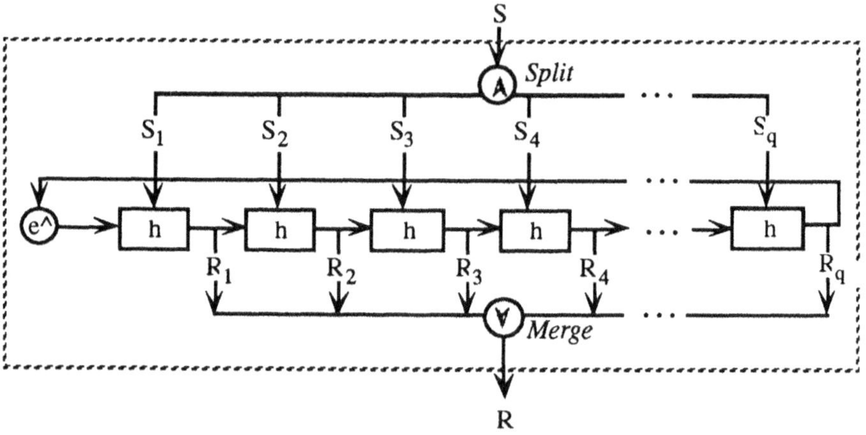


Unfortunately, this is a pretty bad solution, because the parallelism is only superficial. The feedback loop actually prohibits a truly parallel activity of the various processors. Therefore we apply a further transformation: The processors are duplicated, and the feedback loop is transferred to the second set of processors. This leads to the network which is used in the rule given below. The formal derivation of this rule follows the derivation for the function `psums` in sections 2.2 and 2.3. All we have to do, is replace '+' by an arbitrary *associative* function h.

*Proof*: We start from the above net:

$$R_1 = h\star(S_1, e\hat{}R_q)$$
$$R_2 = h\star(S_2, R_1)$$
$$\dots$$
$$R_q = h\star(S_q, R_{q-1})$$

Now we introduce auxiliary streams (where $\bar{e} \overset{\text{def}}{=} \langle e,e,e,e,\dots\rangle$ is the infinite stream of neutral elements):

$$Q_1 = h\star(S_1, \bar{e}) = S_1$$
$$Q_2 = h\star(S_2, Q_1)$$
$$\dots$$
$$Q_q = h\star(S_q, Q_{q-1})$$

From these equations we can immediately deduce (since h is associative):

$$R_1 = h\star(Q_1, e\hat{}R_q)$$
$$R_2 = h\star(S_2, h\star(Q_1, e\hat{}R_q)) = h\star(h\star(S_2, Q_1), e\hat{}R_q) =$$
$$= h\star(Q_2, e\hat{}R_q)$$

...

```
Rq = h★(Sq,h★(Qq-1,e^Rq)) = h★(h★(Sq,Qq-1),e^Rq) =
   = h★(Qq,e^Rq)
```

*(End of proof)*

| ***Rule 3*** |
|---|

Any function definition meeting the pattern

```
DEF (F S) = (h/)★(inits S),
```

with the properties

```
LAW Associative(h) ∧ Neutral(h,e)
```

is equivalent to

```
DEF (F S) = (Merge∘Net∘Splitq)(S)
DEF Net(S1,...,Sq) = R1,...,Rq
    WHERE Q1 = h★(S1,ē)       R1 = h★(Q1,e^Rq)
          Q2 = h★(S2,Q1)      R2 = h★(Q2,e^Rq)
             ...                 ...
          Qq = h★(S1,Qq-1)    Rq = h★(Qq,e^Rq)
```

The `Net` is illustrated by the following diagram

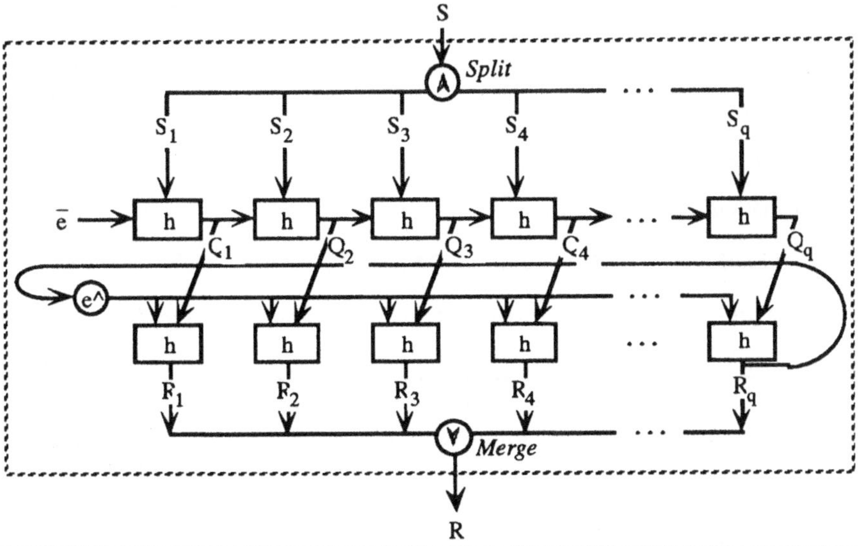

If we analyze the above net from an operational point of view, then we find true parallelism. After a more or less sequential setup phase, where data has to flow from left to right through the upper half, all processors are continuously working.

*Variations on the theme.* We have already seen in the example `psums` that the original specification could be transformed into various recursive solutions. Since recursions of this kind may also occur independently form such a specification, we list the corresponding transformations here as well:

**Rule 4**

A function definition of the kind

```
DEF (F x^S) = (h x)☆(e^(F S)),
```

is equivalent to

```
DEF (F S) = (h/)☆(inits S).
```

This makes rule 3 applicable, provided that `h` is associative and `e` its neutral element.

**Rule 5**

Consider a function definition

```
DEF (F S) = ...
```

with the properties (where `last` denotes the last element of a – finite – stream, and `h` is a suitable operation)

```
LAW last(F S)  = (g S)
LAW (g S^x)    = h((g S),x).
```

This is equivalent to

```
DEF (F S) = (h/)☆(inits S).
```

This makes rule 3 applicable, provided that `h` is associative and `e` its neutral element.

*Finite streams.* The above derivations were given for the case of infinite streams. But it is immediately clear that the corresponding properties hold for any initial section of these streams as well.

```
LAW ((F S) 1..j) = (R 1..j)                    -- for 0≤j≤n
    WHERE
    (R 1..j) = ((h☆(S,c^R)) 1..j)
             = h☆((S 1..j),((e^R) 1..j))
             = h☆((S 1..j),e^(R 1..j-1))
```

The realization by a suitable network remains essentially unchanged, when $n$ is a multiple of the repetition factor $q$. Otherwise, some of the right streams are one element shorter than the left streams, which, however, does not do any harm. Again, these technical details are well within the realm of optimizing compilers.

## 3.3 Another Feedback Situation

In the previous rules we dealt with recursive situations of the kind

```
DEF (F x^S) = (h x)☆(c^(F S)),
LAW (F S) = R  WHERE  R = h☆(S,c^R).
```

The above recursions can be rewritten into the following form, where `(g x)` $\stackrel{\text{def}}{=}$ `(h x c)`:

```
DEF (F x^S) = (g x)^((h x)☆(F S)),
LAW (F x^S) = R  WHERE  R = (g x)^(h☆(S,R)).
```

Now we want to consider the related situation (where `g` often will be the identity function)

```
DEF (F x^S) = (g x)^(F (h∘g x)☆S).
```

Here the '☆'-operator is not applied to the *result* of the recursive call of `F` but rather to its *argument*.

The computations of one incarnation of this function obviously can be performed by a network of the following kind (illustrated for a stream of length eight):

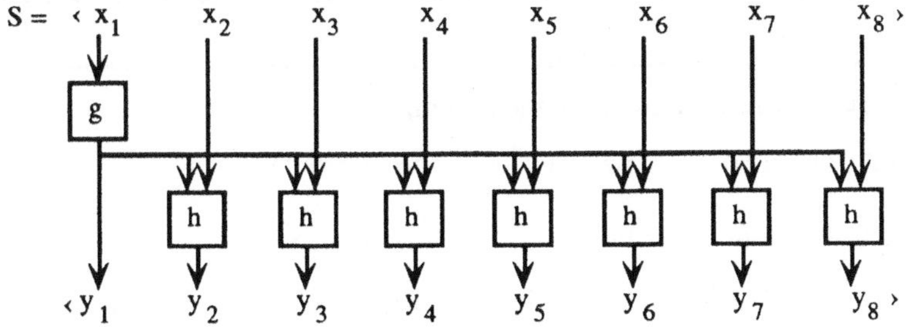


Now the output of the `g`-box is the first element of the result stream, the outputs of the `h`-boxes are fed into the next incarnation of the function `F`, and so on. This is illustrated by the following diagram:

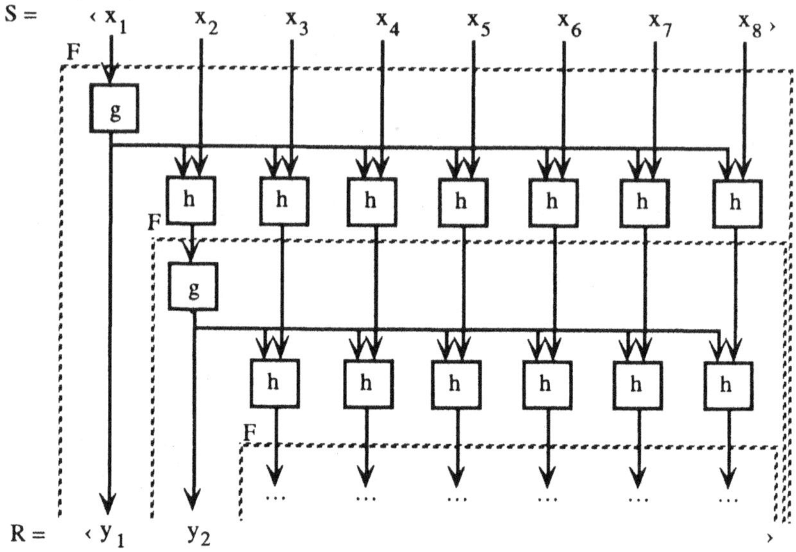


Let us now consider the i-th component of the result stream R. By a simple induction it is immediately seen that this component results from the evaluation of the expression

```
(R i) = (g ∘ h(R i-1) ∘ ... ∘ h(R 2) ∘ h(R 1)) (S i)
```

This expression is computed by a function `H` that takes as input the initial section `(R 1..i-1)` of the result stream `R` and the element `(S i)`.

```
DEF (H x empty) = (g x)
DEF (H x  r^R ) = (H (h r x) R)
```

This constitutes our desired network. (And since its graphical representation is less readable than the corresponding textual representation, we only give the latter in the rule below.)

| ***Rule 6*** |
| --- |

Any function definition meeting the pattern

```
DEF (F x^S) = (g x)^(F (h∘g x)☆S),
```

yields the stream `R` defined by

```
LAW (F S) = R  WHERE  (R i) = (H (S i) (R 1..i-1)).
```

The auxiliary function `H` is defined by

```
DEF (H x empty) = (g x)
DEF (H x  r^R ) = (H (h r x) R)
```

This stream R is generated by the net

```
DEF (F S) =
      LET n = length S IN
     NET
      AGENTS H i = H (S i)                           [i = 1..n]
      CONFIG (H 1).in = empty
             (H i).in = (H 1).out^...^(H i-1).out    [i>1]
      OUTPUT <(H 1).out, ..., (H n).out>
```

*What we have achieved by now is the following situation: We have a couple of general rules that allow us to convert certain high-level expressions into nets of parallel processes. This is, of course, trivial for simple applications of the 'apply-to-all' operator as exemplified in rules 1 and 2; but our schemes also work in cases that involve recursively defined streams, thus leading to feedback loops in the networks, as exemplified in rules 3 - 6.*

# 4 Gaussian Elimination

*Let us now consider an application for the above rule 6. It turns out that the Gaussian elimination is a good example.*

The essential task of Gaussian elimination can be described as follows[1]: Given a matrix A, find a lower triangular matrix L (with diagonal 1) and an upper triangular matrix U such that

$$A = L \cdot U.$$

Actually, this is not exactly true, because there are some transpositions involved due to the so-called "pivot elements". So the actual equation reads

$$P \cdot A = L \cdot U,$$

where $P = P_{n-1} \cdot \ldots \cdot P_1$ is the product of permutation matrices, which are used to exchange certain rows.

Due to the special form of L and U we can actually merge them into one rectangular matrix. We denote this merging by the operator '&':

---

1 We concentrate here on the problem of finding the decomposition into triangular matrices. Algorithms for solving the resulting triangular systems of equations are given e.g. by Fernandez et al. [1991].

```
              u u u u u u u u u
              l u u u u u u u u
              l l u u u u u u u
              l l l u u u u u u
L & U    =    l l l l u u u u u
              l l l l l u u u u
              l l l l l l u u u
              l l l l l l l u u
              l l l l l l l l u
```

*Note*: When working with matrices, we use the following notations:

| | |
|---|---|
| `(A i j)` | element i,j of matrix A |
| `A→i` | i-th row of matrix A |
| `A↓j` | j-th column of matrix A |
| `a^A` | concatenation of vector a to matrix A (depending on the context a is a new row or column) |

## 4.1 A Recursive Solution

*The following deriviation is a condensed version of the development presented by Pepper and Möller [20].*

***Milestone 1:** A recursive solution.* Our first goal is to come up with a recursive solution. Therefore we consider one major step of Gaussian elimination, together with its recursive continuation. Each of these major steps consists of two parts.

1. The first part is the search of a so-called "pivot element". This is the largest element in the first column. The corresponding row *p* then has to be exchanged with the first row. We denote this operation by `(A swap p)`.

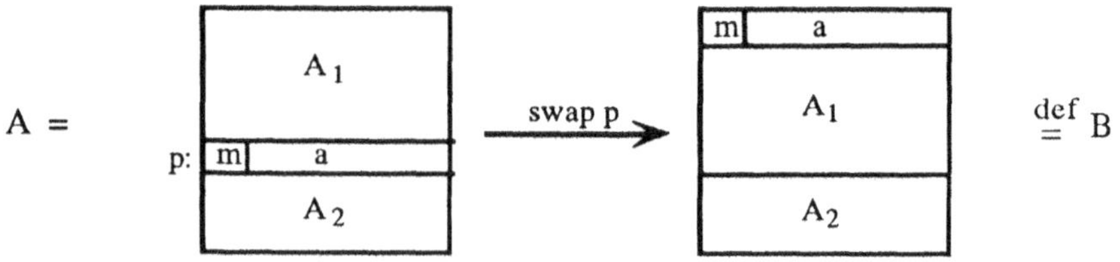


2. The second part comprises the essential work of each step: We produce the first column of `L` and the first row of `U`. The following derivation describes, how this is done. Let `B=(A swap p)` be given. We want to find

```
D = (Gauss B) = (L & U)
```

such that `L·U=B` (actually, `L·U=P·B` for some permutation matrix `P`). We consider the following partitioning of our matrices:

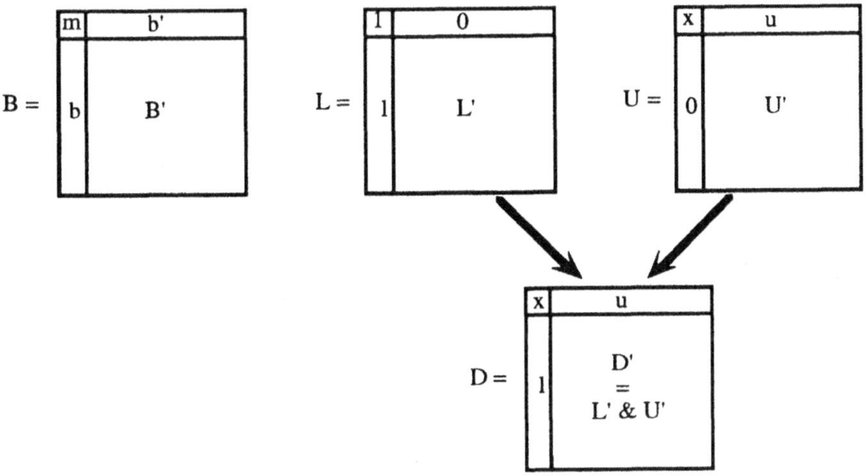


By calculating the matrix product `B=(L·U)` we obtain the following equalities for the fragments of `D`:

```
x       = m
u       = b'
l       = (div x)☆b          [since b  = l·x]
L'·U'   = -☆(B',l·u)         [since B' = +☆(l·u, L'·U')]
```

From this we obtain finally

```
D'  = Gauss(-☆(B',l·u))   [since (L'& U') =
                                         Gauss(L'·U')]
```

Summing up, one major step of Gaussian elimination shall perform the following transition:

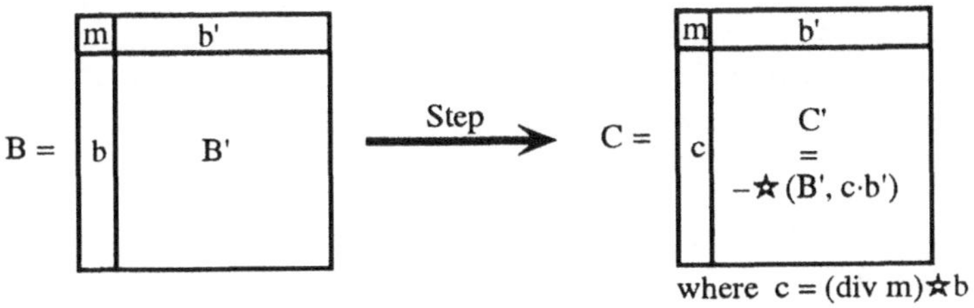


After this step we issue a recursive call `D'=(Gauss C')` for the submatrix `C'`.

## 4.2 Partitioning of the Data Space

Now we want to describe the step from `A` to `C` in such a way that the process can be easily parallelized. We consider a proper partitioning of the data space on the result side, that is, a partitioning of `C`.

***Milestone 2:*** *Towards a column-oriented calculation.* As we had seen in the previous example, the clue to parallelization lies in the use of high-level operators. Therefore we now want to represent the above calculations in a more compact form. We have a choice between a row-oriented approach and a column-oriented approach. Since rows are permuted due to pivoting, the latter approach is preferable. The above equations can be reformulated in a column-oriented fashion as follows:

- The first column of `C` has as its top element the maximal element of the first column of `A` (the pivot element); the remainder of the column (denoted by the operation `rest`) is obtained by permuting the first column of `A` correspondingly and by dividing all elements by the pivot element. All this is abbreviated as an operation `Phi`:

```
(C↓1) = (B 1 1)^c
            where c = (div m)☆(rest B↓1)
                    = (div m)☆(rest(A↓1 swap p))
                  m = (A p 1)
      = (A p 1)^(div (A p 1))☆(rest(A↓1 swap p))
  def
   =  Phi(A↓1)
```

- The remainder of each column of C is obtained by essentially subtracting the corresponding element of the first column from it, multiplied by the first element of the column. That is, we subtract multiples of the first row from all other rows. This process is codified in an operation called `Tau`:

```
(C↓j) = (B 1 j)^(-☆(B'↓j, ((B 1 j) mult)☆c))
      = (A p j)^(-☆(rest((A swap p)↓j),
                      ((A p j) mult)☆rest(C↓1))
  def
   =  Tau(C↓1,A↓j)
```

*In the sequel we are only interested in the structural dependencies; the technical details of the various computations are not relvant for our further considerations. Therefore we hide all these technicalities within the functions* `Phi` *and* `Tau` *as introduced above. Thus, we consider from now on the equations*

```
(C↓1)  =  Phi(A↓1)
(C↓j)  =  Tau(C↓1,A↓j)         for j > 1
```

So our function `Gauss` takes as its input a tuple of columns, where each column is in turn a tuple of values, and it produces as its output again a tuple of columns. Using our by now standard apply-to-all operator '☆', this can be compactly defined as follows.

```
2. Column-oriented recursive solution
-------------------------------------
FUN Gauss: seq[col] → seq[col]

DEF (Gauss empty) = empty

DEF (Gauss a^A)   = LET
                        c = (Phi a)
                    IN
                        c^(Gauss (Tau c)☆A)
```

(Actually, this version may be a little too abstract, because it burdens the function `Tau` also with the splitting of the columns into an already fully processed upper part, where `Tau` is just the identity, and a lower part, on which the actual processing has to be performed. But since we want to concentrate in the sequel on structural aspects, we abstract away as many technical details as possible)

***Milestone 3:*** This definition is an exact instance of rule 6 from section 3.3 above. Therefore we immediately obtain the pertinent network.

```
3. Network solution
-------------------
FUN Gauss: seq[col] → seq[col]

DEF Gauss A =
     LET n = length A IN
    NET
     AGENTS G j = G (A↓j)                          [j = 1..n]
     CONFIG (G 1).in = empty
            (G j).in = (G 1).out ^…^ (G j-1).out   [j>1]
     OUTPUT <(G 1).out, ..., (G n).out>
-------------------
FUN G: col → stream[col] → stream[col]

DEF (G a empty) = (Phi a)

DEF (G a c^C)   = (G (Tau c a) C)
```

## 4.2 Scheduling For $q<<N$ Processors

If the number of available processors would be just $N$ (i.e. the size of A), then our development would be completed, because we could assign one processor to each of the agents in the network. If the number of available processors would even exceed $N$, then we might cluster them into groups and let each group handle the task of one agent. This way, we could even implement some of the fine-grain parallelism inherent in the functions `Tau` and `Phi`. But the most likely situation will be that the number $q$ of available processors is much smaller than the size $N$ of the matrix. Therefore we briefly discuss this variant in the sequel.

We obtain again an application of Brent's theorem (cf. section 3.1). That is, we assign to each concrete processor the tasks of several virtual processors. We choose the simplest distribution of the workload here by the following assigmnet:

Processor `Pi`: Tasks `(G i)`, `(G q+i)`, `(G 2q+i)`, ....

This is nothing but another instance of our `Split/Merge` paradigm from section 3. Note, however, that we are now working with streams, the elements of which are whole columns.

Since we have seen the principle before, we omit the technical details here. But we want to at least point out one additional aspect: For the new agents `Pi` there are essentially two options:

1. Either `Pi` performs its associated tasks `(G i)`, `(G q+i)`, `(G 2q+i)`, ... sequentially.
2. Or `Pi` performs the tasks `(G i)`, `(G q+i)`, `(G 2q+i)`, ... in an interleaved mode.

In a shared-memory environment both variants are equally good. Hence, the greater conceptual simplicity gives preference to the first variant. In a distributed-memory environment the second variant has advantages. To see this, consider the first result column. After it is computed, it is broadcasted to all other processes. This means that it has to be kept by the last process until it has computed all its columns except for the last one. This entails a large communication storage.

Based on these observations, the processes `Pi` should perform their associated tasks in an interleaved fashion. There is only one modification of this principle: When on some column the last `Tau` operation has been performed, the concluding `Phi` operation should be performed right away. This makes the result immediately available to the subsequent tasks and thus prohibits unnecessary delays.

*Remark 1*: As can be seen from the discussion by Quinn [25], our solution is not optimal. Nevertheless, we think that it is a good solution for the following reasons: Firstly, it is not far from optimal. Secondly, in the (slightly) better solutions each column is treated by several processors - which causes considerable communication costs in the case of distributed memory. Finally, our solution is simpler to program, which eases verification as well as adaption to related problems. *(End of remark)*

*Remark 2*: During the Workshop it turned out that our derivation fits nicely into the work of A. Gerasoulis. The relations that we calculated in section 3.3 (of which the function `Gauss` is an instance) determine a dependency graph that can be fed into Gerasoulis' system. This system will then deduce the network sketched above. (As a matter of fact, it will do slightly better than we did, as is elaborated in Gerasoulis' paper.) This demonstrates that a proper abstraction can lead to situations, where the remaining technical work can be taken over by automated systems. *(End of remark)*

# 5 Matrix Multiplication

There is another standard task for parallelization, viz. the problem of multiplying two matrices. We want to at least briefly sketch, how a solution for this problem can be formally deduced. For reasons to be explained below we are content with square $n \times n$-matrices.

## 5.1 The Milestones of the Development

*We first present an overview of the underlying ideas. As usual, this is done in the form of a succession of milestones, which are sufficiently close to each other such that the overall derivation is intuitively acceptable. In the next subsection we will then discuss the derivation concepts from a more abstract point of view.*

***Milestone 1:*** *Specification of the problem.* The standard specification of matrix multiplication simply says that each element of the result matric C is the product of the corresponding row of A and column of B.

| 1. Initial specification: |
|---|
| FUN ·: matrix × matrix → matrix<br>SPC A·B = C ⇒ (C i j) = (A→i)×(B↓j) |

Here we employ the same matrix notations as in the previous section; in addition, we use

axb $\stackrel{\text{def}}{=}$ add/(mult☆(a,b)) scalar product of a and b

***Milestone 2:*** *Partitioning of the data space.* Our strategy is to distribute the result matrix C over the given processors; and for the time being we assume that there is one processor available for each element of C. This design entails that each processor must see one full row of A and one full column of B. However, in order to allow for a high degree of parallelism we decide to "rotate" both the row and the column through the processors. This idea is illustrated by the following diagrams.

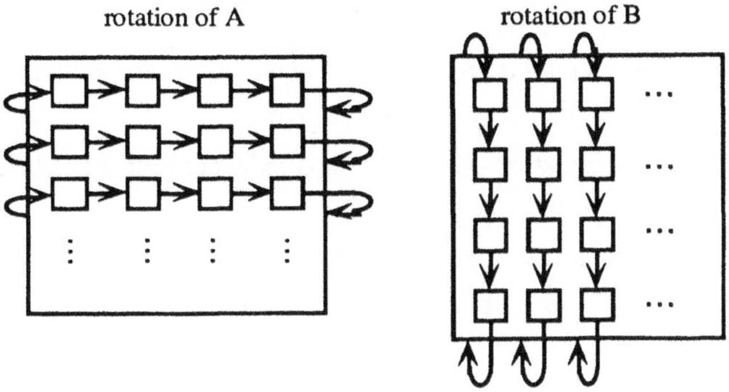

Now we have to turn this intuitive idea into a formal program derivation. First of all, we need an operation `rot` that shifts a vector (i.e. a row or a column); let `x` be an element and `a` be a vector. Then we define (where `last` denotes the last element of a sequence, and `front` denotes the sequence without its last element):

```
DEF rot(a^x) = x^a    that is    rot a = (last a)^(front a).
```

Then we obtain immediately the following property for the scalar product of two vectors `a,b`:

```
LAW a×b = (rot^i a)×(rot^i b).                               [1]
```

On this basis, we decide (after a little exploratory calculation) to let the elements `(C i j)` be computed according to

```
LAW (C i j) = (rot^k A→i) × (rot^k B↓j)
        where k = (i+j) mod n.                               [2]
```

(We assume here that rows and columns are numbered from 0..n–1; but a numbering from 1..n works as well.) From the above property `[1]` we know that `[2]` is correct. Therefore we obtain a correct "solution" as follows:

```
2. A first solution
---------------------------------------------------------------
DEF A·B = ×★CC    WHERE (CC i j) = [rot^k A→i, rot^k B↓j]
                              k = (i+j) mod n
```

In this solution, `CC` is a matrix, the elements of which are pairs consisting of a suitably rotated row of `A` and a suitably rotated column of `B`. And then the scalar product is applied to each such pair in `CC`. This is, of course, a fictitious design, that still needs to be made operational.

***Milestone 3:*** *Introducing the proper data flow.* As mentioned earlier, we want to assign a processor $P_{i,j}$ to each element of the resut matrix `C` (or `CC`, respectively). From the above solution we make the following observation: We only have to ensure that the processor receives the two vectors that constitute the corresponding elements of `CC`. But this follows easily from the following fact:

```
rot^k a = rot(rot^(k-1) a) = last(rot^(k-1) a)^front(rot^(k-1) a)
        = first(rot^k a)^front(rot^(k-1) a)
        = (a k)^front(rot^(k-1) a)
```

This leads to the following recurrence relation between the components of `CC`: Let

```
(CC i j) ≝ [(AA i j),(BB i j)]
```

Then we can derive the equations (where `k-1` is to be computed modulo `n`):

```
(AA i j) = (rot^k A→i)
         = (A i k)^front(rot^(k-1) A→i)
         = (A i k)^front(AA i j-1)
```

```
(BB i j) = (rot^k B↓j)
         = (B k j)^front(rot^{k-1} B↓j)
         = (B k j)^front(BB i-1 j)
```

Hence it suffices that each processor $P_{i,j}$ initially possesses the k-th row element `(A i k)` and the k-th column element `(B k j)`, where `k=(i+j)mod n`. The remainders of the row and column are sent by the neighbouring processors $P_{i-1,j}$ and $P_{i,j-1}$, respectively. This design is captured by the following net implementation:

```
3. Network implementation for matrix multiplication
--------------------------------------------------------------------
FUN ·: matrix × matrix → matrix

DEF A · B =
    LET n = size A IN
       -- Note: all index calculations by means of ⊕
       -- are 'mod n'
       -- the range of i,j is: i = 0..n-1, j = 0..n-1
    NET
      AGENTS (P i j) = P((A i k),(B k j)) where k=(i⊕j)
      CONFIG (P i j).right = (P i j⊕1).left
             (P i j).bot   = (P i⊕1 j).top
      OUTPUT (P i j).res
--------------------------------------------------------------------
-- main function

FUN P: (num × num)
        → left:stream[num] × top:stream[num]
         → right:stream[num]× bot:stream[num] × res:num

DEF P(a0,b0)(left,top).right = front(a0^left)
                       .bot   = front(b0^top)
                       .res   = (a0^left)×(b0^top)
```

This is a dataflow design that is still adaptable to a synchronized behaviour in the style of SIMD-machines or to the asynchronous behaviour of MIMD-machines.

***Milestone 4:*** *Considering distributed memory.* The above solution does not fit the paradigm of distributed memory, because each processor only has to hold one element from each of the matrices `A` and `B`, and an intermediate result. On the other hand, the assumption that we have $n^2$ processors available is in general unrealistic. The solution is, however, obvious, and therefore we will only sketch it here briefly. Suppose that we have $q^2$ processors available (for $q \ll n$); then we *tile* the given matrix correspondingly:

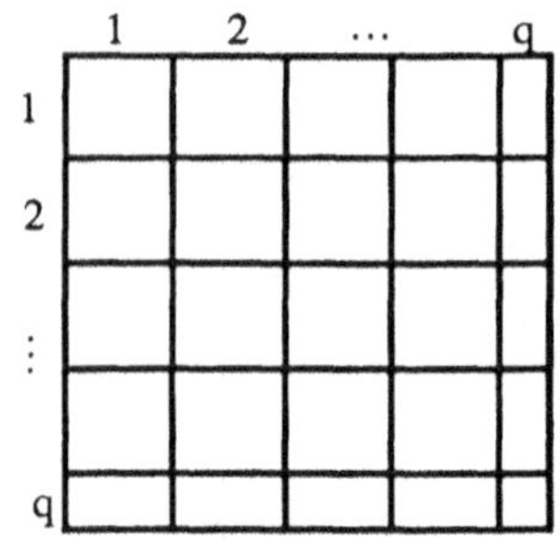


To this variant we then apply our above algorithm. The "only" difference now is that in the place of single numbers we encounter complete submatrices; this merely affects the type `stream[num]` and the scalar product '×' in the above definition of `P`. The type now becomes `stream[matrix]`, and in the scalar product `a×b=add/(mult☆(a,b))` the operations `add` and `mult` now stand for matrix addition and matrix multiplication. All this can be easily achieved in ML-like languages by using polymorphic functions. Therefore we need not invest additional development efforts here.

*Discussion*

- *Now our design has the effect that each processor holds two such submatrices, does all the computations that need to be done with this data, and then passes them on to its neighbours. Hence, we obtain the minimal communication overhead. Moreover, the communication pattern is optimal for most real machines: Only neighbouring processors have to communicate, and the communication is unidirectional.*
- *The final derivation step leading to stage 4 clearly is another instance of Brent's theorem (cf. section 3.1).*

## 5.2 Morphisms From Data Structures To Net Structures

*The above example demonstrates most clearly a principle that was also present in some of the other examples: The form of the underlying data structures determines the structure of the network of processes. This phenomenon is well-known under the buzzword* ***systolic algorithms****. We will now try to get a transformational access to this paradigm. (Even though we do not yet have a fully worked-out theory for these ideas, it may still be worthwhile to sketch the general scheme.)*

In all our examples, we work with data structures such as sequences, trees, or matrices. Let us denote this data structure by $\sigma[\alpha]$, where $\alpha$ stands for the underlying basic data type such as `num` in `matrix[num]`. For such data structures `S`:$\sigma[\alpha]$ we presuppose selection operations denoted by `(S i)` as in the case of sequences or `(S i j)` as in the case of matrices.

Let us now consider the above example of the matrix multiplication; there we were able to deduce for the data structure CC some kind of recurrence relation, and this relation determined the layout of the dataflow network. More abstractly speaking, suppose that the data structure S under consideration obeys a relation of the kind

```
(S i) = Ψ[x, (S j), (S k)]
```

where x, j, and k depend on S and i. Then the computation of

```
h☆S
```

is achieved by a network of the kind

```
NET: σ[α]
   AGENTS (P i) = (P x)
   CONFIG (P i).in1 = (P j).out
          (P i).in2 = (P k).out
   OUTPUT (P i).res
DEF (P x)(in1,in2).out = Ψ[x,in1,in2]
                  .res = h(in1,in2)
```

A comparison with the example for the matrix multiplication will help to elucidate the principal way of proceeding. There the decisive relationship is

```
(CC i j) = Ψ[(A i k),(B k j),(CC i j-1),(CC i-1 j)]
```

This determines two facts: Firstly, the process $P_{i,j}$ initially needs the elements (A i k) and (B k j) in its local memory. Secondly, there is a data flow from $P_{i,j-1}$ and $P_{i-1,j}$ to $P_{i,j}$.

A similar construction would have been possible for the example psums. If we use here the data structure definition

```
(S i) =def sum ((inits B) i)
```

then we obtain the relationship

```
(S i) = (S i-1)+(B i)
      = Ψ[(B i),(S i-1)].
```

This relationship yields the essence of the network in milestone 7 of section 2.2.

*Even though this outline still is a bit vague and sketchy, it already indicates, how our derivations may possibly be lifted to a higher level of abstraction. Based on such morphisms between data and network structure we could concentrate on the derivation of recurrence relations on the purely functional level. The corresponding stream equations – that still are derived individually for each transformation rule in the preceeding sections – would then follow automatically.*

# 6 Warshall's Algorithm

*As an example for a graph algorithm we consider the problem of minimal paths in a graph and the standard solution given by Warshall. Since this algorithms comes again quite close to our previous matrix algorithms, we will keep the treatment very brief and sketchy, initially following the derivation of Pepper and Möller [20].*

***Milestone 1:*** *Initial specification.* We are given a directed graph the edges of which are labelled by "distances", denoted as `(dist i j)`; for simplicity we set `(dist i j)`$=\infty$ when there is no edge from `i` to `j`. As usual, a path is a sequence of nodes $p=\langle i_1, \ldots, i_n\rangle$ such that each pair $i_k$ and $i_{k+1}$ is connected by an edge. The *length* of such a path is the sum of all the edges. On this basis, the problem can be specified as follows

| 1. Initial specification |
|---|
| FUN MinDist: node × node → nat |
| DEF MinDist(i,j) = min{ length p \| (ispath p i j) } |

***Milestone 2:*** *Initial recursive solution.* We employ the following idea: The nodes of the graph are successively coloured black, and at each stage we only allow paths, the inner nodes of which are black. If `B` represents the set of black nodes, we immediately obtain the following properties (where '↓' denotes the minimum of two numbers):

| 2. Recursive solution |
|---|
| FUN MD: set[node] → node × node → nat |
| DEF MD(∅)(i,j) = (dist i j) |
| DEF MD(B∪{a})(i,j) = MD(B)(i,j)↓(MD(B)(i,a)+MD(B)(a,j)) |
| LAW MinDist(i,j) = MD(AllNodes G)(i,j) |

The idea of this solution is evident: If we blacken the node `a`, then we have for any pair of nodes `i,j` the following situation: If their shortest black connection does not go through `a`, nothing changes for them; otherwise there must have been black connections from `i` to `a` and from `a` to `j` before. (The details are worked out by Pepper and Möller [20].)

The additional law simply states, how the initial function `MinDist` is implemented by the new function MD.

***Milestone 3:*** *Matrix-based solution.* The above algorithm exhibits an expensive recursion structure. This can be simplified by the introduction of matrices, where

```
TYPE matrix = node × node → nat;
```

that is, matrices are just used to represent the labelled edges. Hence, we may use the initial function `dist` as initial matrix. So the above definition is simply rewritten by introducing suitable abbreviations:

```
3. Recursive solution in matrix notation

FUN MD: set[node] → matrix

DEF MD(∅) = dist
DEF MD(B∪{a}) = h(M,a)☆M
      WHERE M = MD(B)
            h(M,a)(i,j) = (M i j) ↓ ((M i a) + (M a j))
```

By standard transformations – as listed e.g. by Bauer and Wössner [4] – this recursion can be converted into a simple loop, and `matrix` is a mapping that can be represented by a finite data structure.

***Milestone 4:*** *Partitioning of the data space.* As usual, we have to come up with a suitable partitioning of the data space to prepare the parallelization. In our case, a tiling of the matrix `M` appears to be the best choice (as suggested by Gomm et al. [14]). This is in accordance with the computational dependencies of the above algorithm, as it is illustrated by the following diagram.

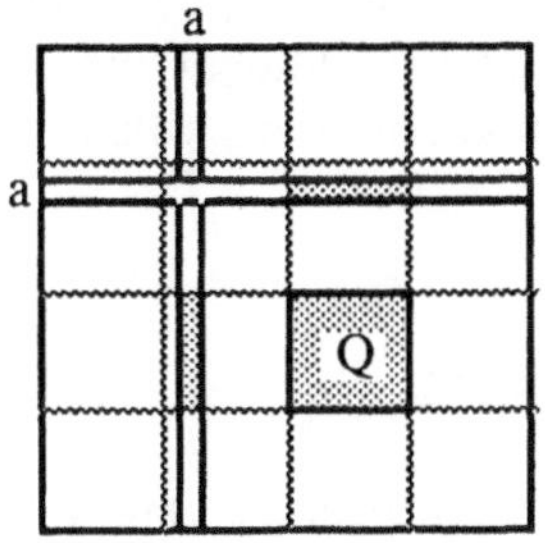


If we now apply the idea from the previous section 5.2, then we obtain the following recurrence relations:

```
Q = (M i j) = Ψ[(M i j),(M i a),(M a j)]
```

The morphism between data structure and network structure then determines that

- each processor $P_{i,j}$ has to hold the submatrix `(M i j)` in its local memory;
- there is a dataflow from $P_{i,a}$ and $P_{a,j}$ to $P_{i,j}$.

*Since the programming of this design proceeds very much along the lines of our previous examples, we refrain from going into the technical details here.*

# 7 Parallel Parsing

The problem of parallel parsing has received considerable interest, as can be seen from the collection of papers assembled by op den Akker et al. [18]. In spite of this widespread interest, the situation is not very satisfactory when looked upon with the eyes of a practitioneer. Most results are useless for one of the following two reasons:

- Optimality results for general parsing mostly are of the kind exemplified by Theorem 4.1 of Gibbons and Rytter [13]: "Every context-free language can be recognized in $O(\log^2 n)$ time using $n^6$ processors on a PRAM." In a setting, where $n$ is in the order of several thousands, this result is at best of academic interest.
- More reasonable results are of the kind exemplified by Theorem 4.5 in the same book: "Every bracket language can be parsed in $O(\log n)$ time using $n/\log n$ processors on a PRAM." (This result has been derived by Bar-On and Vishkin [3].) Unfortunately, these results refer to extremely small classes of languages, such as pure paranthesis languages or even regular languages.

Therefore we take the humble programmer's approach here and perform a very straightforward parallelization of standard sequential parsing, presupposing a fixed number $q$ of available processors with distributed memory. In doing so, we also refrain from employing sequential preprocessing, as it is suggested e.g. by Baccelli and Fleury [2].

We should mention, however, that in spite of the conceptual simplicity of our approach it is just a matter of a few additional transformations to turn our algorithm into the aforementioned optimal solutions, when the languages are appropriately restricted.

## 7.1 The Domain Theory

For our purposes we take a very abstract view of "parsing". For more detailed accounts of the underlying techniques and concepts we refer to the standard textbooks. Through parsing we convert a given string, i.e. a sequence of items

$$\langle a_1, a_2, a_3, \ldots, a_n \rangle$$

into a single item, viz. the so-called parse tree that reflects the inherent grammatical structure of the original string. This conversion proceeds gradually through a series of intermediate strings. Therefore, our "items" can be both tokens and trees. (The tokens are the initial input symbols, the trees result from the parsing of suitable string fragments.)

The basic transition step from one string to the next string can be described as follows: The current string is partitioned into two parts:

$$s = p \downarrow r$$

where the substring p has already been analysed as far as possible, and the substring r yet has to be considered. Hence, p is a "forest" of trees, and r is a sequence of input tokens:

```
                 p                     ↓              r
s =  | forest (sequence of trees)     | sequence of tokens |
```

In this situation there are two possibilities: Either a reduction rule is applicable; this means that p ends in a "handle" h, which can be reduced to a tree t:

```
                           transform
s = q++h ↓ r                  ➠            s' = q^t ↓ r.
```

Or no rule is applicable, which means that the "focus" is shifted ahead over the next input token a:

```
                           transform
s = p ↓ a^r                   ➠            s' = p^a ↓ r.
```

This is the general principle of parsing. The main idea of the so-called LR techniques lies in the form of decision making: Should we perform a shift or a reduction? The details of how these decisions are made do not concern us here[1]; we rather combine all these operations into a single function transform:

```
1. General description of (LR-)parsing

FUN Parse: string → tree

DEF Parse(<t>↓empty) = t
DEF Parse(p↓r)       = Parse(transform(p↓r))

-- auxiliary function:
FUN transform: string → string
SPC transform(s) = «shift or reduce»
```

Note that the function transform encapsulates all the information of the traditional LR-tables, including error handling. But this level of abstraction is acceptable for our purposes.

## 7.2 Towards Parallel Parsing

At first sight this concept seems to be strictly sequential, since it is always the left context p (together with the first token a of the right context r) which decides the next action to be performed; this left context is codified in a so-called *state*. However, in most grammars there are tokens which always lead to the same state, independent of the left context. (They possess a uniform column in the traditional LR-tables.) Such tokens typically are major keywords such as 'TYPE', 'PROC', 'FUN', 'IF', 'WHILE', etc., but – depending on the language design – they may

1 The details of this view of parsing are worked out by Pepper [24].

also be parantheses, colons etc. (These tokens also play a central role for recovering after the detection of errors in the input string.)

*Remark 1*: Now we can see, why the aforementioned "bracket languages" exhibit such a nice behaviour: They only consist of symbols which possess the desired independence from the left context. This demonstrates that our approach is a true generalization of the concepts presented for these simplistic languages. *(End of remark)*

*Remark 2*: We might generalize this idea by associating to each token the set of its possible states. During shift operations these sets could be constrained until a singleton set occurs, in which case the normal parsing could start. But we will not pursue this variant any further here, because its associated overhead may well outweigh its potential gains. *(End of remark)*

Now our function `Parse` has to be generalized appropriately: It does no longer necessarily start at the left side of a given input string, which it then reduces to a single tree, but it may also operate on an arbitrary inner substring, which it shall reduce as much as possible. This entails only a minor modification:

```
2. Partial (LR-)parsing

FUN Parse: string → string
DEF Parse(p↓empty) = p
DEF Parse(p↓r)   = Parse(transform(p↓r))
```

It is evident that the overall parsing process is not affected, when some inner substring has already been partially reduced; that is:

$$\mathtt{Parse}(s_1\mathtt{++Parse}(s_2)\mathtt{++}s_3) = \mathtt{Parse}(s_1\mathtt{++}s_2\mathtt{++}s_3)$$

By induction, the last equation can be generalized to

$$\mathtt{Parse}(s_1\mathtt{++}s_2\mathtt{++}\ldots\mathtt{++}s_q) =$$

$$\mathtt{Parse}(s_1\mathtt{++Parse}(s_2\mathtt{++Parse}(\ldots\mathtt{++Parse}(s_q)\ldots)))$$

This equation is the clue to our parallelization.

*Note*: This idea requires a minor extension of the LR tables. Traditionally, the lead symbols of the right context, the so-called lookahead symbols, can only be terminals. In our setting they can be nonterminals as well.

## 7.3 Partitioning of the Data Space

As usual, our first concern is an appropriate partitioning of the data space. But by contrast to our numerical examples, there is no a-priori separation available. This is immediately seen from the following argument: The best performance would be achieved, if all processors would cope with subtrees (of the final parse tree) that have approximately the same size. The structure of the subtree is, however, only known at the end of the computation. Hence, we can at best make good initial guesses.

Lacking better guidelines, we simply split the initial string into sections of equal length. (At the end of our derivation we will see that this decision should be slightly modified.) If there are *q* processors available, we partition the given string of tokens into q fragments:

$$s = s_1 ++ s_2 ++ \ldots ++ s_q,$$

which are of approximately the same size. And we give one such fragment to each processor.

## 7.4 Conversion to a Network of Processes

On this basis, we can now derive the parallel layout for our algorithm. The development essentially relies on the above equation

$$\text{Parse}(s_1{+}{+}s_2{+}{+}\ldots{+}{+}s_q) =$$
$$\text{Parse}(s_1{+}{+}\text{Parse}(s_2{+}{+}\text{Parse}(\ldots{+}{+}\text{Parse}(s_q)\ldots)))$$

Therefore we set up a simple network of processes with the following structure:

If we take the liberty to freely concatenate strings and streams, then we may just write this in the following form:

```
3. Parallel parsing

FUN ParallelParse: string → string
DEF ParallelParse(s1++s2++...++sq) =
     NET
      AGENTS P i = P(si)                     [i = 1..q]
      CONFIG (P i).in = (P i+1).out          [i = 1..q-1]
             (P q).in = empty
      OUTPUT (P 1).out

FUN P: string → in:Stream[item] → out:Stream[item]
DEF P(s)(in) = Parse(s++in)
```

The remaining details of implementing the interplay between the actual reduction operations and the communication operations is then a standard technical procedure, which is well within the realm of compilers.

This principle can be cast into a general rule:

**Rule 7**

Any function definition meeting the pattern

```
DEF (F S) = (++°F)/(Split S),
```

and obeying the property

```
LAW F(x++y) = F(x++F(y))
```

is equivalent to the network

```
DEF (F S) = (Net°Split_q)(S)
DEF Net(S_1,...,S_q) = R_1  WHERE R_1 = (P S_1 R_2)
                                  R_2 = (P S_2 R_3)
                                   ...
                                  R_q = (P S_q empty)
FUN P: seq[α] → stream[α] → stream[α]
DEF (P S in) = (F S++in)
```

The Net is illustrated by the following diagram

*Discussion*

- *The above design indicates a flaw in our initial data distribution. It can be seen that all processors send their unfinished work to their respective left neighbours. (In the best case, this is just one tree, in the worst case it is the complete and unchanged original substring.) Therefore, the size of the initial substrings* $s_1, \ldots, s_q$ *should increase from left to right. However, the best choice of sizes depends on both the grammar and the concrete input string. Therefore, one can at best look for good heuristics here.*
- *This also shows that a general complexity analysis is not possible here, because it highly depends on the the properties of the concrete grammar under consideration. In the best case we achieve parsing in $O(n/q)$ time. This is, for instance, the case for the aforementioned nice and harmless "bracket languages". In the worst case we are stuck with $O(n)$ time, because no inner processor can do any reductions, thus transferring all the workload to processor $P_1$.*

# 8 On the Reuse of Ideas

*To round off the experimental thoughts of this paper we want to sketch, how the same derivation idea can be applied in various situations. This is, of course, the heart of any kind of programming knowledge, as it is possessed by professionals and taught to students. But we want to emphasize here another point of view: This reuse does not only occur on the informal level of intuitive understanding, but can be applied as a quite formal and technical tool.*

To illustrate our point we consider a technique that often is referred to as *"pointer jumping"*. (This choice of terminology already indicates, how biased most treatments of this idea are by low-level implementation details.) We strive for a more abstract view that will allow us to reuse the concept in different situations.

*Note: In the sequel we only to point out, where the concept could be applied. It is not our goal to actually carry the developments through.*

*Pointer jumping* often is exemplified by the problem of list ranking: The distance of every element from the end of the list shall be computed. In our sequence terminology this means "length of sequence – position of element", where the length of the sequence is initially not known. This problem is easily recognized as a special case of the prefix sums from section 2: Let all list elements be 1 and build the prefix sums of the reversed sequence. Therefore we do not look at list ranking itself, but rather reconsider briefly the problem of the prefix sums.

## 8.1 Odd/Even Splitting For Prefix Sums

The whole idea relies on a binary data space partitioning, that is, on a classical divide-and-conquer approach. Let $S = \langle x_1, \ldots, x_n \rangle$ be a sequence. Then we have

$$(\mathrm{Split}_2\ S) = [S_1, S_2] \quad \text{with} \quad S_1 = \langle x_1, x_3, x_5, x_7, \ldots \rangle$$
$$S_2 = \langle x_2, x_4, x_6, x_8, \ldots \rangle$$

$$(\mathrm{Merge}\ \langle S_1, S_2 \rangle) = S.$$

Now consider the pointwise additions of $S_1$ and $S_2$:

$$Q = +\star(S_1, 0\hat{\ }S_2) \qquad R = +\star(S_1, S_2)$$

with the understanding that in case of different lengths the remainder of the larger sequence is copied identically.

*Example*: Let $S = \langle 1\ 2\ 3\ 4\ 5\ 6\ 7\ 8\ 9 \rangle$. Then:

```
  S1 = 1   3   5   7   9          S1 =  1   3   5   7   9
0^S2 = 0   2   4   6   8          S2 =  2   4   6   8
______________________________   ______________________________

   Q = 1  2+3 4+5 6+7 8+9          R = 1+2 3+4 5+6 7+8  9
```

Note that the first elements of Q and R, respectively, already yield the first two elements of (psums S). Now we apply the same procedure recursively to Q and R:

```
Q1 =  1    4+5  8+9
R1 =  1+2  5+6
Q2 =  2+3  6+7
R2 =  3+4  7+8
```

From these we obtain (where e.g. 2...5 stands for 2+3+4+5):

```
U = +☆(Q1,0^Q2) =  1    2...5  6...9
V = +☆(R1,0^R2) =  1...2  3...6  7...8
W = +☆(Q1,  Q2) =  1...3  4...7  8...9
X = +☆(R1,  R2) =  1...4  5...8
```

Now the lead elements already yield the first four elements of (psums S). These first steps shall suffice to illustrate the following facts:

- By a straightforward induction it is shown that this process finally yields (psums S).
- We can regard all the sequences in each step of this process as a matrix. Then initially S is a (1,N)-matrix. In the next step Q and R together form a (2,N/2)-matrix, and so on, until we arrive at the final (N,1)-matrix. Hence, we operate with a constant overall data size.
- Given $N$ processors, each step is performed in constant time. Then the overall time complexity is O(log $N$) steps. (Using Brent's theorem again, the same time complexity can be obtained using $N$/log $N$ processors.)

The formal definition of this algorithm may be given as follows:

```
FUN P: seq[seq[num]] → seq[seq[num]]
DEF (P M) = LET M' = ([g,h]∘Split2)☆M
                IN  (proj1☆M')++(proj2☆M')
```

with the definitions

```
g(S1,S2)  = +☆(S1,0^S2)        proj1(A,B) = A
[g,h](x)  = <(g x),(h x)>
h(S1,S2)  = +☆(S1,S2)          proj2(A,B) = B.
```

Obviously, this is the style of equations to which the methods from the previous sections apply. (Therefore we do not go into further details here.)

## 8.2 Odd/Even Splitting For Polynomial Interpolation

The Aitken-Neville algorithm for polynomial interpolation is based on the following recursive equations (where E is a suitable numeric expression and y a vector of start values)

```
(P 0 i) = (y i)
(P k i) = E[(P k-1 i), (P k-1 i+1)]
```

As outlined by Pepper [22], the computation of this definition can be parallelized using an odd/even splitting. To see this, consider the two successive vectors

$$p \stackrel{def}{=} (P\ k)$$

$$p' \stackrel{def}{=} (P\ k+1)$$

Then we have the relationships

```
p' = E☆(front p, rest p)
```

(because `(p' i) = E[(p i), (p i+1)]`)

Now consider the splittings

$r,s = (Split_2\ p)$ $\quad$ $r',s' = (Split_2\ p')$

This induces the relationships

```
r' = E☆(r, s)
s' = E☆(r, rest s)
```

Again, we have thus reached a state, where the techniques from the previous sections are applicable.

## 8.3 Odd/Even Splitting For the Fast Fourier Tranform

Finally, it is nice to observe that the rather intricate computation of the Fast Fourier Transformation also falls into this category. The task here is to evaluate a polynomial

$$p(x) \stackrel{def}{=} a_0+a_1x+ a_2x^2 + \dots +a_{n-1}x^{n-1}$$

at $n$ distinct points $r_0,\dots,r_{n-1}$. As has been elaborated by Pepper and Möller [20], the solutions of this problem can be specified as

$$fft(a) = \langle y_0,\dots,y_{n-1}\rangle \quad \text{where} \quad y_0 = a \times r^0$$

$$\dots$$

$$y_{n-1} = a \times r^{n-1}$$

where a is the vector of coefficients $\langle a_0, a_1, a_2, \dots, a_{n-1}\rangle$ and r is a special vector based on the so-called "n-th roots of unity". (The details are of no concern here.) Certain algebraic properties of the n-th roots of unity r then allow us to apply the odd/even splitting in order to convert this system of equations into the form

```
(fft a) = +☆(fft(even a),*☆(r,fft(odd a))) ++
          -☆(fft(even a),*☆(r,fft(odd a))).
```

The common subexpressions `fft(even a)` and `fft(odd a)` are responsible for the complexity *O(n log n)*. And the high-level operators enable a parallel evaluation.

# 9 Summary

In the natural sciences, researchers perform experiments in order to extract common principles that will ultimately be codified in theories. And it is the purpose of these theories that they can be applied later on to analogous situations, allowing predictions about the outcomes to be expected.

In computer science, we may perform case studies in order to extract common principles that will ultimately be codified in "programming theories". And it is the purpose of these theories that they can be applied later on to analogous situations. However, since computer science is closer to engineering than to the natural sciences, our theories are not used to predict the behaviour of existing systems, but rather to guide us in the construction of new systems. This way, the experience gained by researchers is made available to program engineers, who have to build the "real" systems, where the well-known time and cost constraints do not allow them to treat every software project as a new scientific challenge.

What we have tried in this paper is to perform a variety of case studies, concentrating on the issue of correctness rather than on complexity aspects. Our goal was to give systematic derivations of parallel algorithms, such that their correctness is ensured by construction. In doing so, we wanted to stay as close as possible to the large body of strategies and techniques that has been acquired in the area of sequential-program development over the last years. And it turned out that this is indeed feasible. The greater part of our derivations could be done by using standard procedures. There was only need for a very few extensions, in particular:

- Techniques for expressing appropriate data space partitionings, such as
  - for sequences: partitioning into sequences of sequences;
  - for matrices: partitioning into rows or columns, or tilings into submatrices;

  etc.
- Conversion of high-level operators such as *apply-to-all* or *reduce* into nets of communicating parallel processes.

Of course, there still is considerable technical work to be done. For instance, several of our examples exhibit a behaviour that is only clumsily captured by our high-level operators: There is some kind of feedback, which leads to a combination of the operators with some additional recursive definitions of sequences, vectors etc. It is not obvious, how these situations should be abstracted into new kinds of operators.

Darlington et al. [10] study various so-called *skeletons*, that is, special higher-order functions, that are similar to the kinds of patterns discussed here. For these skeletons there exist translations into different parallel machine architectures. This means that we can indeed develop algorithms on a very high and abstract level and then compile them into code for (parallel or sequential) machines. However, these studies are only in their beginning phase, and it is not yet clear, how far this paradigm will reach, and what efforts will be required for actually programming in this style. Nevertheless, the diversity of examples treated in the previous sections indicates that any parallel program that can be developed using traditional techniques can also be developed using our techniques. And the experience gained with sequential programs is also backed by our case studies for the parallel case: The abstract and high-level development is much more elegant and secure.

Since our case studies aimed at methodological insights, we were quite liberal with respect to notations and programming models; in particular, it was not our goal to invent new syntaxes for parallel programming. As a consequence we also had to state our "programming theories" in a very liberal fashion. But experience shows that the codification into the formalism of, say, program transformation does not yield new insights; it is only a mandatory prerequisite, when one aims at semi-automatic assistance by transformation or verification systems.

In summary, our experiment revealed that the time is ripe for a more engineering-oriented approach also in the area of parallel programming. We must get away from the situation, where each tiny program is treated like a precious little gem, valuable, hard to find, and costly. It is time that programs, be they sequential or parallel, can be created in a systematic way as normal work products of professional engineers, rather than being the topic of PhD theses and scientific papers.

## References

Abbreviations:

PPOPP'91 = 3rd ACM SIGPLAN Symp. n Principles & Practice of Parallel Programming, SIGPLAN Notices 26:7, July 1991.

EDMCC2 = Bode, A. (ed.): *Distributed Memory Computing*. Proc. 2nd European Conf. EDMCC2, Munich, April 1991, Lecture Notes in Computer Science 487. Berlin: Springer 1991.

-------------

[1] Aiken, A., Nicolau, A.: Perfect Pipelining: A New Loop Parallelization Technique. In: Proc. ESOP. Berlin: Springer 1988.

[2] Baccelli, F., Fleury, T.: *On Parsing Arithmetic Expressions in a Multi-Processing Environment.* Acta Informatica 17, 1982, 287-310.

[3] Bar-On, I., Vishkin, U.: *Optimal Parallel Generation of a Computation Tree Form.* ACM Trans. Prog. Lang. and Systems 7:2, 1985, 348-357.

[4] Bauer, F.L., Wössner, H.: *Algorithmic Language And Program Development.* Berlin: Springer 1982.

[5] Bird, R.: *Lectures on Constructive Funtional Programming.* In: Broy, M. (ed.): Constructive Methods in Computing Science. Proc. Int. Summer School, Berlin: Springer 1989.

[6] Bode, A. (ed.): *Distributed Memory Computing*. Proc. 2nd European Conf. EDMCC2, Munich, April 1991, Lecture Notes in Computer Science 487. Berlin: Springer 1991.

[7] Broy, M.: *A Theory For Nondeterminism, Parallelism, Communications, and Concurrency*. Theor. Comp. Sc. 45, 1-61 (1986).

[8] Cole, M.: *Algorithmic Skeletons: Structured Management of Parallel Computation*. Pitman/MIT Press, 1989.

[9] Cormen, T.H., Leiserson, C.E:, Rivest, R.L.: *Introduction to Algorithms*. Cambridge: MIT Press and New York: McGraw-Hill, 1991.

[10] Darlington, J., Field, A.J., Harrison, P.G., Kelly, P.H.J., While,R.L., Wu, Q.: *Parallel Programming Using Skeleton Functions*. Techn. Rep. Dept. of Computing, Imperial College, London, May 26, 1992.

[11] Fernandez, A., Llaberia, J.M., Navarro, J.J., Valero-Garcia, M.: *Interleaving Partitions of Systolic Algorithms for Programming Distributed Memory Multiprocessors*. In EDMCC2, 1991.

[12] Gallager, R.G., Humblet, P.A., Spira, P.M.: *A Distributed Algorithm for Minimum-Weight Spanning Trees*. ACM TOPLAS 5:1 (1983) 66-77.

[13] Gibbons, A., Rytter, W.: *Efficient Parallel Algorithms*. Cambridge: Cambridge University Press1988.

[14] Gomm, D., Heckner, M., Lange, K.-J., Riedle, G.: *On the Design of Parallel Programs for Machines with Distributed Memory*. In: EDMCC2, 1991. 381-391

[15] Gries, D.: *The Science of Programming*. New York: Springer 1981.

[16] Knapp, E.: *An Exercise in the Formal Derivation of Parallel Programs:Maximum Flow in Graphs*. ACM TOPLAS 12:2 (1990) 203-223.

[17] Nicolau, A., Wang, H.: *Optimal Schedules for Parallel Prefix Computation with Bounded Resources*. In: PPOPP'91, 1-10.

[18] op den Akker, R., Albas, H., Nijholt, A., Oude Luttighuis, P.: *An Annotaded Bibliography on Parallel Parsing*. Universiteit Twente, faculteit der informatica, Memoranda Informatica 89-67, Dec. 1989.

[19] Partsch, H.: *Specification And Transformation of Programs*. Berlin: Springer 1990.

[20] Pepper, P., Möller, B.: *Programming With (Finite) Mappings*. In: M. Broy (ed.): Informatik und Mathematik. Berlin: Springer 1991. 381-405.

[21] Pepper, P., Schulte, W.: Some Experiments on the Optimal Compilation of Applicative Programs. In: M. Bettaz (ed.): Proc. First Maghrebinian Seminar on Sofware Engineering and Artifiacial Intelligence, Constantine, Algeria, Sept. 1989.

[22] Pepper, P.: *Specification Languages and Program Transformation*. In: Reid, J.K. (ed.): Relationship between Numerical Computation and Programming Languages. Proc. IFIP WG 2.5 Conf., Boulder 1981. Amsterdam, North-Holland 1982, 331-346.

[23] Pepper, P.: *Literate program derivation: A case study.* Broy, M., Wirsing, M. (Eds.): Methodik des Programmierens. Lecture Notes in Computer Science 544, Berlin: Springer 1991, 101-124.

[24] Pepper, P.: *Grundlagen des Übersetzerbaus.* Course manuscript, Techn. Univ. Berlin, 1991.

[25] Quinn, M.J.: *Designing Efficient Algorithms for Parallel Computers.* New York: McGraw-Hill 1987.

[26] Schulte, W., Grieskamp, W.: *Generating Efficient Portable Code for a Strict Applicative Language.* To appear in Proc. Phoenix Seminar and Workshop on Declarative Programming, Hohritt, Germany, Nov. 1991.

[27] Tel, G., Tan, R.B., van Leeuwen, J.: *The Derivation of Graph Marking Algorithms from Distributed Termination Detection Protocols.* Science of Comp. Progr. 10 (1988) 107-137.

[28] Wile, D.: *Local Formalisms: Widening the Spectrum of Wide-Spectrum Languages.* In: Meertens, L.G.L.T. (ed.): Proc. IFIP TC2 Working Conf. on Program Specification and Transformation, Bad Tölz. Noth-Holland 1986, 459-481.

[29] Yang, J.A., Choo,Y.: *Parallel-Program Transformation Using a Metalanguage.* In: PPOPP'91, 11-20.

# 2

# Derivation of Parallel Sorting Algorithms

Douglas R. Smith[1]
email: smith@condor.kestrel.edu
Kestrel Institute
3260 Hillview Avenue
Palo Alto, California 94304 USA

**Abstract**

Parallel algorithms can be derived from formal problem specifications by applying a sequence of transformations that embody information about algorithms, data structures, and optimization techniques. The KIDS system provides automated support for this approach to algorithm design. This paper carries through the salient parts of a formal derivation for a well-known parallel sorting algorithm - Batcher's Even-Odd sort. The main difficulty lies in building up the problem domain theory within which the algorithm is inferred.

## 1 Introduction

This paper is intended as an introduction to formal and automated design of parallel algorithms. The level of formality is somewhat lessened in order to concentrate on the main issues. We derive Batcher's Odd-Even sort [2] and discuss the derivation of several other well-known parallel sorting algorithms.

Algorithms can be treated as a highly optimized composition of information about the problem being solved, algorithm paradigms, data structures, target architectures, and so on. An attempt to provide automated support for algorithm design must be based on a formal model of the composition process and

1. *representation of problem domain knowledge* - expressing the basic and derived concepts of the problem and the laws for reasoning about them. We formalize knowledge about a particular application domain as a parameterized *domain theory.*

2. *representation of programming knowledge* - we also use theories to capture knowledge of algorithms and data structures. The logical concept of interpretation between theories is the basis for applying programming knowledge in the form of theories [3, 10, 12].

Most, if not all, sorting algorithms can be derived as interpretations of the divide-and-conquer paradigm. Accordingly, we present a simplified divide-and-conquer theory and show how it can be applied to design the sort algorithms mentioned above.

There are a variety of reasons for turning to a derivational approach to algorithm design. First, a derivation is structured *proof of correctness*, so a derivational approach is in accordance with modern programming methodology

[1] This research was supported in part by the Office of Naval Research under Grant N00014-90-J-1733 and in part by Air Force Office of Scientific Research under Contract F49620-91-C-0073.

that insists that programs and proofs be developed at the same time. Second, the compositional view provides an *explanation* of an algorithm in terms that are common to many algorithms. This description shows the commonalities between algorithms and how a small collection of general principles suffice to generate a large variety of algorithms. All too often, the published explanation of an algorithm is just a post-hoc proof of correctness that sheds little light on the process of inventing the algorithm in the first place. Such proofs are too specific to the algorithm and use overly general proof techniques, such as induction. The reader may wish to compare our derivation with the presentation of Even-Odd sort in textbooks such as [1, 5]. Third, derivations often come in families – the design decisions that are dependent on the target language and architecture can be separated out. This allows *retargeting* a abstract algorithm for a problem to a variety of concrete programs in different languages for different machines. Finally, *automated support* can be provided for formal derivations. The machine handles many of the lower-level details, freeing the designer to concentrate on developing the problem domain theory and making high-level design decisions.

# 2 KIDS Model of Design

The Kestrel Interactive Development System (KIDS) has served as a testbed for our experiments in automated program derivation [9]. The user typically goes through the following steps in using KIDS. We do not claim this to be a complete model of software development; however, this model is supported in KIDS and has been used to design and optimize over 60 algorithms. Applications areas have included scheduling, combinatorial design, sorting and searching, computational geometry, pattern matching, and linear and nonlinear programming.

1. *Develop a domain theory* – The user builds up a domain theory by defining appropriate types and operations. The user also provides laws that allow high-level reasoning about the defined operations. Our experience has been that laws describing the preservation of properties under various operations provide most of the laws that are needed to support design and optimization. In particular, distributive and monotonicity laws have turned out to be so important that KIDS has a theory development component that supports their automated derivation.

2. *Create a specification* – The user enters a specification stated in terms of the underlying domain theory.

3. *Apply a design tactic* – The user selects an algorithm design tactic from a menu and applies it to a specification. Currently KIDS has tactics for simple problem reduction (reducing a specification to a library routine), divide-and-conquer, global search (binary search, backtrack, branch-and-bound), local search (hillclimbing), and problem reduction generators (dynamic programming and generalized branch-and-bound).

4. *Apply optimizations* – KIDS supports program optimization techniques such as simplification, partial evaluation, finite differencing, and other

transformations. The user selects an optimization method from a menu and applies it by pointing at a program expression.

5. *Apply data type refinements* – The user can select implementations for the high-level data types in the program. Data type refinement rules carry out the details of constructing the implementation [3].

6. *Compile* – The resulting code is compiled to executable form. In a sense, KIDS can be regarded as a front-end to a conventonal compiler.

Actually, the user is free to apply any subset of the KIDS operations in any order – the above sequence is typical of our experiments in algorithm design. In this paper we mainly concentrate on the first three steps.

# 3 Derivation of a Mergesort

## 3.1 Domain Theory for Sorting

Suppose that we wish to sort a collection of objects belonging to some set $\alpha$ that is linearly-ordered under $\leq$. Here is a simple specification of the sorting problem:

$$\begin{array}{l} Sort(x : bag(\alpha) \mid true) \\ returns(\ z : seq(\alpha) \mid x = Seq\text{--}to\text{--}bag(z) \ \wedge \ Ordered(z)\ ) \end{array}$$

*Sort* takes a bag (multiset) $x$ of $\alpha$ objects and returns some sequence $z$ such that the following *output condition* holds: the bag of objects in sequence $z$ is the same as $x$ and $z$ must be ordered under $\leq$. The predicate *true* following the parameter $x$ is called the *input condition* and specifies any constraints on inputs.

In order to support this specification formally, we need a domain theory of sorting that includes the theory of sequences and bags, has the linear-order $\langle \alpha, \leq \rangle$ as a parameter, and defines the concepts of *Seq–to–bag* and *Ordered*. The following parameterized theory accomplishes these ends:

**Theory** $Sorting(\langle \alpha, \leq \rangle : linear - order)$
  **Imports** $integer,\ bag(\alpha), seq(\alpha)$
  **Operations**
    $Ordered : seq(\alpha) \rightarrow Boolean$

  **Axioms**

$$\forall (S : seq(\alpha))\ (Ordered(S) \Leftrightarrow \forall (i)(i \in \{1..length(S) - 1\} \implies S(i) \leq S(i+1)))$$

  **Theorems**

$$\begin{array}{l} Ordered([]) = true \\ \forall (a : \alpha)\ (Ordered([a]) = true) \\ \forall (y_1 : seq(\alpha), y_2 : seq(\alpha)) \\ \quad (Ordered(y_1 +\!\!+ y_2) \Leftrightarrow Ordered(y_1) \\ \qquad \wedge\ Seq\text{--}to\text{--}bag(y_1) \leq Seq\text{--}to\text{--}bag(y_2) \\ \qquad \wedge\ Ordered(y_2)) \end{array}$$

**end–theory**

*Sorting* theory imports *integer*, *bag*, and *sequence* theory. Sequences are constructed via [] (empty sequence), [a] (singleton sequence), and $A{+\!\!+}B$ (concatenation). For example,

$$[1,2,3]{+\!\!+}[4,5,6] = [1,2,3,4,5,6].$$

Several parallel sorting algorithms are based on an alternative set of constructors which use interleaving in place of concatenation: the *ilv* operator

$$[1,2,3] \ ilv \ [4,5,6] = [1,4,2,5,3,6]$$

interleaves the elements of its arguments. We assume that the arguments to *ilv* have the same length, typically denoted $n$, and that it is defined by

$$A \ ilv \ B = C \Leftrightarrow \forall(i)(i \in \{1..n\} \implies C_{2i-1} = A_i \ \wedge \ C_{2i} = B_i).$$

In Section 4 we develop some of the theory of sequences based on the *ilv* constructor.

Bags have an analogous set of constructors: $\{\!\!\{\}\!\!\}$ (empty bag), $\{\!\!\{a\}\!\!\}$ (singleton bag), and $A \uplus B$ (associative and commutative bag union). The operator *Seq-to-bag* coerces sequences to bags by forgetting the ordering implicit in the sequence. *Seq-to-bag* obeys the following distributive laws:

$$Seq\text{-}to\text{-}bag([]) = \{\!\!\{\}\!\!\}$$

$$\forall(a:\alpha) \ Seq\text{-}to\text{-}bag([a]) = \{\!\!\{a\}\!\!\}$$

$$\forall(y_1 : seq(\alpha), y_2 : seq(\alpha))$$
$$Seq\text{-}to\text{-}bag(y_1{+\!\!+}y_2) = Seq\text{-}to\text{-}bag(y_1) \uplus Seq\text{-}to\text{-}bag(y_1)$$

$$\forall(y_1 : seq(\alpha), y_2 : seq(\alpha))$$
$$Seq\text{-}to\text{-}bag(y_1 \ ilv \ y_2) = Seq\text{-}to\text{-}bag(y_1) \uplus Seq\text{-}to\text{-}bag(y_1)$$

In the sequel we will omit universal quantifiers whenever it is possible to simplify the presentation without sacrificing clarity.

## 3.2 Divide-and-Conquer Theory

Most sorting algorithms are based on the divide-and-conquer paradigm: If the input is primitive then a solution is obtained directly, by simple code. Otherwise a solution is obtained by decomposing the input into parts, independently solving the parts, then composing the results. Program termination is guaranteed by requiring that decomposition is monotonic with respect to a suitable well-founded ordering. In this paper we focus on divide-and-conquer algorithms that have the following general form:

$DC(x_0 : D \mid I(x_0))$
$returns(\ z : R \mid O(x_0, z))$
$= \textbf{if} \ Primitive(x_0)$
$\quad \textbf{then} \ Directly\text{-}Solve(x_0)$
$\quad \textbf{else} \ let \ \langle x_1, x_2 \rangle = Decompose(x_0)$
$\qquad Compose(DC(x_1), \ DC(x_2))$

We refer to *Decompose* as a decomposition operator, *Compose* as a composition operator, *Primitive* as a control predicate, and *Directly–Solve* as a primitive operator.

The essence of a divide-and-conquer algorithm can be presented via a *reduction diagram*:

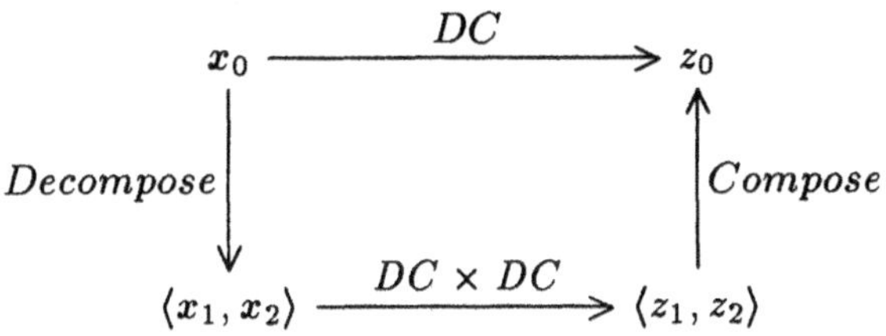


which should be read as follows. Given input $x_0$, an acceptable solution $z_0$ can be found by decomposing $x_0$ into two subproblems $x_1$ and $x_2$, solving these subproblem recursively yielding solutions $z_1$ and $z_2$ respectively, and then composing $z_1$ and $z_2$ to form $z_0$.

In the derivations of this paper we will usually ignore the primitive predicate and *Directly-Solve* operator – the interesting design work lies in calculating compatible pairs of *Decompose* and *Compose* operators.

The following mergesort program is an instance of this scheme:

```
MSort(b0 : bag(integer))
    returns( z : seq(α) | x = Seq–to–bag(z) ∧ Ordered(z) )
    = if size(b0) ≤ 1
        then b0
        else let ⟨b1, b2⟩ = Split(b0)
                Merge(MSort(b1), MSort(b2))
```

Here *Split* decomposes a bag into two subbags of roughly equal size and *Merge* composes two sorted sequences to form a sorted sequence.

The characteristic that subproblems are solved independently gives the divide-and-conquer notion its great potential in parallel environments. Another aspect of divide-and-conquer is that the recursive decomposition can often be performed implicitly, thereby enabling a purely bottom-up computation. For example, in the Mergesort algorithm, the only reason for the recursive splitting is to control the order of composition (merging) of sorted subproblem solutions. However the pattern of merging is easily determined at design-time and leads to the usual binary tree computation pattern.

To express the essence of divide-and-conquer, we define a divide-and-conquer theory comprised of various sorts, function, predicates, and axioms that assure that the above scheme correctly solves a given problem. A simplified divide-and-conquer theory is as follows (for more details see [8, 11]):

**Theory** *Divide–and–Conquer*

**Sorts** $D, R$ — domain and range of a problem

**Operations**

| | |
|---|---|
| $I : D \rightarrow Boolean$ | input condition |
| $O : D \times R \rightarrow Boolean$ | output condition |
| $primitive : D \rightarrow Boolean$ | control predicate |
| $O_{Decompose} : D \times D \times D \rightarrow Boolean$ | output condition for Decompose |
| $O_{Compose} : R \times R \times R \rightarrow Boolean$ | output condition for Compose |
| $\succ : D \times D \rightarrow Boolean$ | well-founded order |

**Soundness Axiom**

$$O_{Decompose}(x_0, x_1, x_2) \wedge O(x_1, z_1) \wedge O(x_2, z_2) \wedge O_{Compose}(z_0, z_1, z_2) \implies O(x_0, z_0)$$

...

**end–theory**

The intuitive meaning of the Soundness Axiom is that if input $x_0$ decomposes into a pair of subproblems $\langle x_1, x_2\rangle$, and $z_1$ and $z_2$ are solutions to subproblems $x_1$ and $x_2$ respectively, and furthermore solutions $z_1$ and $z_2$ can be composed to form solution $z_0$, then $z_0$ is guaranteed to be a solution to input $x_0$. There are other axioms that are required: well-foundedness conditions on $\succ$ and *admissability conditions* that assure that Decompose and Compose can be refined to total functions over their domains. We ignore these in order to concentrate on the essentials of the design process.

The main difficulty in designing an instance of the divide-and-conquer scheme for a particular problem lies in constructing decomposition and composition operators that work together. The following is a simplified version of a tactic in [8].

1. Choose a simple decomposition operator and well-founded order.

2. Derive the control predicate based on the conditions under which the decomposition operator preserves the well-founded order and produces legal subproblems.

3. Derive the input and output conditions of the composition operator using the Soundness Axiom of divide-and-conquer theory.

4. Design an algorithm for the composition operator.

5. Design an algorithm for the primitive operator.

Mergesort is derived by choosing $\uplus^{-1}$ as a simple (nondeterministic) decomposition operator. A specification for the well-known merge operation is derived using the Soundness Axiom.

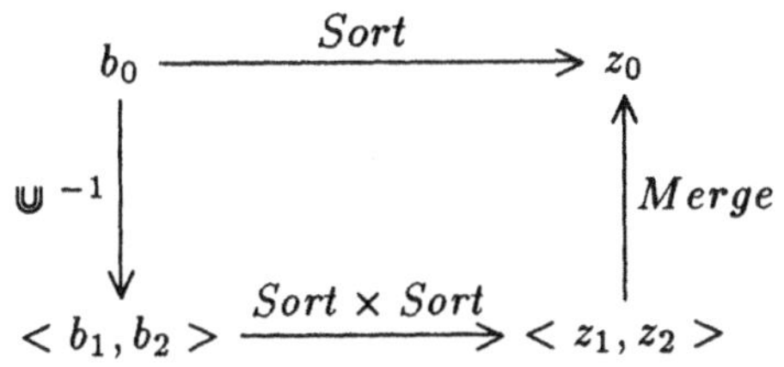


A similar tactic based on choosing a simple composition operator and then solving for the decomposition operator is also presented in [8]. This tactic can be used to derive selection sort and quicksort-like algorithms.

Deriving the output condition of the composition operator is the most challenging step and bears further explanation. The *Soundness Axiom* of divide-and-conquer theory relates the output conditions of the subalgorithms to the output condition of the whole divide-and-conquer algorithm:

$$\begin{array}{l} O_{Decompose}(x_0, x_1, x_2) \\ \wedge\ O(x_1, z_1)\ \wedge\ O(x_2, z_2) \\ \wedge\ O_{Compose}(z_0, z_1, z_2) \\ \Longrightarrow\ O(x_0, z_0) \end{array}$$

For design purposes this constraint can be treated as having three unknowns: $O$, $O_{Decompose}$, and $O_{Compose}$. Given $O$ from the original specification, we supply an expression for $O_{Decompose}$ then reason backwards from the consequent to an expression over the program variables $z_0$, $z_1$, and $z_2$. This derived expression is taken as the output condition of *Compose*.

Returning to Mergesort, suppose that we choose $\uplus^{-1}$ as a simple decomposition operator. To proceed with the tactic, we instantiate the Soundness Axiom with the following substitutions

$$\begin{array}{rcl} O_{Decompose} & \mapsto & \lambda(b_0, b_1, b_2)\ b_0 = b_1 \uplus b_2 \\ O & \mapsto & \lambda(b, z)\ b = \mathit{Seq\text{--}to\text{--}bag}(z)\ \wedge\ \mathit{Ordered}(z) \end{array}$$

yielding

$$\begin{array}{l} b_0 = b_1 \uplus b_2 \\ \wedge\ b_1 = \mathit{Seq\text{--}to\text{--}bag}(z_1)\ \wedge\ \mathit{Ordered}(z_1) \\ \wedge\ b_2 = \mathit{Seq\text{--}to\text{--}bag}(z_2)\ \wedge\ \mathit{Ordered}(z_2) \\ \wedge\ O_{Compose}(z_0, z_1, z_2) \\ \Longrightarrow\ b_0 = \mathit{Seq\text{--}to\text{--}bag}(z_0)\ \wedge\ \mathit{Ordered}(z_0) \end{array}$$

To derive $O_{Compose}(z_0, z_1, z_2)$ we reason backwards from the consequent $b_0 = \mathit{Seq\text{--}to\text{--}bag}(z_0)\ \wedge\ \mathit{Ordered}(z_0)$ toward a sufficient condition expressed over the variables $\{z_0, z_1, z_2\}$ modulo the assumptions of the antecedent:

$b_0 = \mathit{Seq\text{–}to\text{–}bag}(z_0) \;\wedge\; \mathit{Ordered}(z_0)$

$$\Longleftrightarrow \qquad \text{using assumption } b_0 = b_1 \uplus b_2$$

$$b_1 \uplus b_2 = \mathit{Seq\text{–}to\text{–}bag}(z_0) \;\wedge\; \mathit{Ordered}(z_0)$$

$$\Longleftrightarrow \qquad \text{using assumption } b_i = \mathit{Seq\text{–}to\text{–}bag}(z_i),\ i = 1,2$$

$$\mathit{Seq\text{–}to\text{–}bag}(z_1) \uplus \mathit{Seq\text{–}to\text{–}bag}(z_2) = \mathit{Seq\text{–}to\text{–}bag}(z_0) \wedge \mathit{Ordered}(z_0).$$

This last expression is a sufficient condition expressed in terms of the variables $\{z_0, z_1, z_2\}$ and so we take it to be the output condition for *Compose*. In other words, we ensure that the Soundness Axiom holds by taking this expression as a constraint on the behavior of the composition operator.

The input condition to the composition operator is obtained by forward inference from the antecedent of the soundness axiom; here we have the (trivial) consequences $\mathit{Ordered}(z_1)$ and $\mathit{Ordered}(z_2)$. Only consequences expressed in terms of the input variables $z_1$ and $z_2$ are useful.

Thus we have derived a formal specification for *Compose*:

$$\begin{aligned}&\mathit{Merge}(A : \mathit{seq}(\mathit{integer}), B : \mathit{seq}(\mathit{integer}) \mid \mathit{Ordered}(A) \wedge \mathit{Ordered}(B))\\ &\quad \mathit{returns}(\ z : \mathit{seq}(\mathit{integer})\\ &\qquad\qquad \mid \mathit{Seq\text{–}to\text{–}bag}(A) \uplus \mathit{Seq\text{–}to\text{–}bag}(B) = \mathit{Seq\text{–}to\text{–}bag}(z)\\ &\qquad\qquad \wedge \mathit{Ordered}(z)\ ).\end{aligned}$$

Merge is now a derived concept in *Sorting* theory. We later derive laws for it, but now we proceed to design an algorithm to satisfy this specification. The usual sequential algorithm for merging is based on choosing a simple "cons" composition operator and deriving a decomposition operator [8]. However this algorithm is inherently sequential and requires linear time.

# 4 Batcher's Odd-Even Sort

Batcher's Odd-Even sort algorithm [2] is a mergesort algorithm in which the merge operator itself is a divide-and-conquer algorithm. The Odd-Even merge is derived by choosing a simple decomposition operator based on *ilv* and deriving constraints on the composition operator.

Before proceeding with algorithm design we need to develop some of the theory of sequences based on the *ilv* constructor. Generally, we develop a domain theory by deriving laws about the various concepts of the domain. In particular we have found that distributive, monotonicity, and invariance laws provide most of the laws needed to support formal design. This suggests that we develop laws for various sorting concepts, such as *Seq-to-bag* and *Ordered*. From Section 3 we have

**Theorem 1.** *Distributing Seq-to-bag over sequence constructors.*

1.1. $\mathit{Seq\text{–}to\text{–}bag}([]) = \{\!\{\}\!\}$
1.2. $\mathit{Seq\text{–}to\text{–}bag}([a]) = \{\!\{a\}\!\}$
1.3. $\mathit{Seq\text{–}to\text{–}bag}(S_1\ \mathit{ilv}\ S_2) = \mathit{Seq\text{–}to\text{–}bag}(S_1) \uplus \mathit{Seq\text{–}to\text{–}bag}(S_2)$

It is not obvious how to distribute *Ordered* over *ilv* , so we try to derive it. In this derivation let $n$ denote the length of both $A$ and $B$.

*Ordered(A ilv B)*

$\iff$ by definition of *Ordered*

$$\forall(i)(i \in \{1..2n-1\} \implies (A\ ilv\ B)_i \leq (A\ ilv\ B)_{i+1})$$

$\iff$ change of index

$$\forall(j)(j \in \{1..n\} \implies (A\ ilv\ B)_{2j-1} \leq (A\ ilv\ B)_{2j})$$
$$\wedge\ \forall(j)(j \in \{1..n-1\} \implies (A\ ilv\ B)_{2j} \leq (A\ ilv\ B)_{2j+1})$$

$\iff$ by definition of *ilv*

$$\forall(j)(j \in \{1..n\} \implies A_j \leq B_j)$$
$$\wedge\ \forall(j)(j \in \{1..n-1\} \implies B_j \leq A_{j+1}).$$

These last two conjuncts are similar in form and suggest the need for a new concept definition and perhaps new notation. Suppose we define $A \leq^* B$ *iff* $A_i \leq B_i$ for $i \in \{1 \ldots n\}$. This allows us to express the first conjunct as $A \leq^* B$, but then we cannot quite express the second concept – we need to generalize to allow an offset in the comparison:

**Definition 1.** A pair of sequences $A$ and $B$ of length $n$ are *pairwise-ordered with offset k*, written $A \leq^*_k B$, iff $A_i \leq B_{i+k}$ for $i \in \{1 \ldots n-k\}$.

Then the derivation above yields the following simple law

**Theorem 2.** *Conditions under which an interleaved sequence is Ordered.*
For all sequences $A$, $B$,
*Ordered(A ilv B)* $\iff A \leq^*_0 B \wedge B \leq^*_1 A$.

Note that this definition provides a proper generalization of the notion of orderedness:

**Theorem 3.** *Ordered as a diagonal specialization of* $\leq^*_i$.
For all sequences $S$,
*Ordered(S)* $\iff S \leq^*_1 S$

Other laws are easily derived:

**Theorem 4.** *Transitivity of* $\leq^*_i$.
For all sequences $A$, $B$, $C$ of equal length and integers $i$ and $j$,
$A \leq^*_i B \wedge B \leq^*_j C \implies A \leq^*_{i+j} C$

As a simple consequence we have

**Corollary 1.** *Only Ordered sequences interleave to form Ordered sequences.*
For all sequences $A$, $B$,
*Ordered(A ilv B)* $\implies$ *Ordered(A)* $\wedge$ *Ordered(B)*.

Proof:

$Ordered(A \; ilv \; B)$

$$\Longleftrightarrow \quad \text{by Theorem 2}$$

$$A \leq_0^* B \wedge B \leq_1^* A$$

$$\Longrightarrow \quad \text{applying Theorem 4 twice}$$

$$A \leq_1^* A \wedge B \leq_1^* B$$

$$\Longleftrightarrow \quad \text{by Theorem 3}$$

$$Ordered(A) \wedge Ordered(B). \; \Box$$

**Theorem 5.** *Monotonicity of $\leq_i^*$ with respect to merging.*
For all sequences $A_1$, $A_2$, $B_1$, and $B_2$ and integers $i$,
$$A_1 \leq_i^* A_2 \wedge B_1 \leq_i^* B_2 \Longrightarrow Merge(A_1, B_1) \leq_{2i}^* Merge(A_2, B_2)$$

We can apply the basic sort operation $sort2(x, y) = \langle min(x, y), max(x, y)\rangle$ over parallel sequences, just as we did with the comparator $\leq$.

**Definition 2.** *Pairwise-sort of sequences with offset $k$.*
Define $sort2_k^*(A, B) = \langle A', B'\rangle$ such that
*(1)* for $i \leq k$, $B'_i = B_i$
*(2)* for $i = 1, \ldots, n-k$, $\langle A'_i, B'_{i+k}\rangle = sort2(A_i, B_{i+k})$
*(3)* for $i > n-k$, $A'_i = A_i$

For example, $sort2_1^*([2,3,8,9],[0,1,4,5]) = \langle [1,3,5,9],[0,2,4,8]\rangle$. Laws for $sort2_k^*$ can be developed:

**Theorem 6.** *$sort2_k^*$ establishes $\leq_k^*$.*
For all sequences $A$, $B$, $A'$, and $B'$, and integer $k$,
$sort2_k^*(A, B) = \langle A', B'\rangle \Longrightarrow A' \leq_k^* B'$.

This theorem is a trivial consequence of the definition of $sort2_k^*$. The following theorems give conditions under which important properties of the domain theory ($\leq_i^*$, *Ordered*) are preserved under under the $sort2_k^*$ operation. They can be proved using straightforward analysis of cases.

**Theorem 7.** *Ordered is invariant under $sort2_k^*$.*
For all sequences $A$, $B$ and integer $k$,
$Ordered(A) \wedge Ordered(B) \wedge sort2_k^*(A, B) = \langle A', B'\rangle$
$\Longrightarrow Ordered(A') \wedge Ordered(B')$

**Theorem 8.** *Invariance of $A \leq_i^* B$ with respect to $sort2_k^*(A, B)$.*
For all sequences $A$, $B$ and integers $i$ and $k$,
$A \leq_i^* B \wedge sort2_k^*(A, B) = \langle A', B'\rangle \Longrightarrow A' \leq_i^* B'$

**Theorem 9.** *Invariance of $A \leq_i^* B$ with respect to $sort2_k^*(B, A)$.*
For all sequences $A$, $B$ and $0 \leq i < k$,
$A \leq_{i+k}^* A \wedge B \leq_{i+k}^* B \wedge A \leq_i^* B \wedge B \leq_{i+2k}^* A \wedge sort2_k^*(B, A) = \langle B', A'\rangle$
$\Longrightarrow A' \leq_i^* B'$

With these concepts and laws in hand, we can proceed to derive Batcher's Odd-Even mergesort. It can be derived simply by choosing to decompose the inputs to Merge by uninterleaving them.

$$\begin{array}{ccc}
\langle A_0, B_0 \rangle & \xrightarrow{\quad Merge \quad} & S_0 : seq(integer) \\
\Big\downarrow ilv^{-2} & & \Big\uparrow ? \\
\langle \langle A_1, B_1 \rangle, \langle A_2, B_2 \rangle \rangle & \xrightarrow{\quad Merge \times Merge \quad} & \langle S_1, S_2 \rangle
\end{array}$$

where $ilv^{-2}$ means $A_0 = A_1 \ ilv \ A_2$ and $B_0 = B_1 \ ilv \ B_2$. Note how this decomposition operator creates subproblems of roughly the same size which provides good opportunities for parallel computation. Note also that this decomposition operator must ensure that the subproblems $\langle A_1, B_1 \rangle$ and $\langle A_2, B_2 \rangle$ satisfy the input conditions of *Merge*. This property is assured by Corollary 1.

We proceed by instantiating the Soundness Axiom as before:

$$\begin{array}{l}
A_0 = A_1 \ ilv \ A_2 \ \wedge \ Ordered(A_0) \\
\wedge \ B_0 = B_1 \ ilv \ B_2 \ \wedge \ Ordered(B_0) \\
\wedge \ Seq\text{–}to\text{–}bag(S_1) = Seq\text{–}to\text{–}bag(A_1) \uplus Seq\text{–}to\text{–}bag(B_2) \ \wedge \ Ordered(S_1) \\
\wedge \ Seq\text{–}to\text{–}bag(S_2) = Seq\text{–}to\text{–}bag(A_2) \uplus Seq\text{–}to\text{–}bag(B_2) \ \wedge \ Ordered(S_2) \\
\wedge \ O_{Compose}(S_0, S_1, S_2) \\
\Longrightarrow \ Seq\text{–}to\text{–}bag(S_0) = Seq\text{–}to\text{–}bag(A_0) \uplus Seq\text{–}to\text{–}bag(B_0) \\
\qquad \wedge \ Ordered(S_0)
\end{array}$$

Constraints on $O_{Compose}$ are derived as follows:

$$Seq\text{–}to\text{–}bag(S_0) = Seq\text{–}to\text{–}bag(A_0) \uplus Seq\text{–}to\text{–}bag(B_0) \ \wedge \ Ordered(S_0)$$

$\Longleftrightarrow$ by assumption

$$\begin{array}{l}
Seq\text{–}to\text{–}bag(S_0) = Seq\text{–}to\text{–}bag(A_1 \ ilv \ A_2) \\
\qquad \uplus \ Seq\text{–}to\text{–}bag(B_1 \ ilv \ B_2) \\
\wedge \ Ordered(S_0)
\end{array}$$

$\Longleftrightarrow$ distributing *Seq–to–bag* over *ilv*

$$\begin{array}{l}
Seq\text{–}to\text{–}bag(S_0) = Seq\text{–}to\text{–}bag(A_1) \uplus Seq\text{–}to\text{–}bag(A_2) \\
\qquad \uplus \ Seq\text{–}to\text{–}bag(B_1) \uplus Seq\text{–}to\text{–}bag(B_2) \\
\wedge \ Ordered(S_0)
\end{array}$$

$\Longleftrightarrow$ by assumption

$$\begin{array}{l}
Seq\text{–}to\text{–}bag(S_0) = Seq\text{–}to\text{–}bag(S_1) \uplus Seq\text{–}to\text{–}bag(S_2) \\
\wedge \ Ordered(S_0).
\end{array}$$

The input conditions on *Merge* are derived by forward inference from the assumptions above:

$$\begin{array}{l} A_0 = A_1 \; ilv \; A_2 \; \wedge \; Ordered(A_0) \\ \wedge \; B_0 = B_1 \; ilv \; B_2 \; \wedge \; Ordered(B_0) \\ \wedge \; Ordered(S_1) \; \wedge \; Ordered(S_2) \end{array}$$

$\Longrightarrow$ distributing *Ordered* over *ilv*

$$\begin{array}{l} A_1 \leq_0^* A_2 \; \wedge \; A_2 \leq_1^* A_1 \\ \wedge \; B_1 \leq_0^* B_2 \; \wedge \; B_2 \leq_1^* B_1 \\ \wedge \; Ordered(S_1) \; \wedge \; Ordered(S_2) \end{array}$$

$\Longrightarrow$ by monotonicity of $\leq_i^*$ with respect to *Merge*

$$\begin{array}{l} S_1 \leq_0^* S_2 \; \wedge \; S_2 \leq_2^* S_1 \\ \wedge \; Ordered(S_1) \; \wedge \; Ordered(S_2). \end{array}$$

Thus we have derived the specification

$$\begin{array}{l} Merge\text{--}Compose(S_1 : seq(integer), S_2 : seq(integer) \\ \qquad \mid S_1 \leq_0^* S_2 \; \wedge \; S_2 \leq_2^* S_1 \; \wedge \; Ordered(S_1) \; \wedge \; Ordered(S_2)) \\ \quad returns( \; S_0 : seq(integer) \\ \qquad \mid Seq\text{--}to\text{--}bag(S_0) = Seq\text{--}to\text{--}bag(S_1) \uplus Seq\text{--}to\text{--}bag(S_2) \\ \qquad \wedge \; Ordered(S_0) \; ). \end{array}$$

How can this specification be satisfied? Theorems 1.3 and 2 suggest *ilv* since it would establish the output conditions of *Merge–Compose*. Theorem 2 requires that we achieve the input condition $S_1 \leq_0^* S_2 \; \wedge \; S_2 \leq_1^* S_1$ first. But Theorem 6 ($sort2_k^*$ establishes $\leq_k^*$) enables us to apply $Sort2_1^*(S_2, S_1)$ in order to achieve the second conjunct. Theorems 7, 8, and 9 ensure that $S_1 \leq_0^* S_2$ remains invariant. So *Merge–Compose* is satisfied by $ilv \cdot sort2_1^*(S_2, S_1)$. The final algorithm in diagram form is

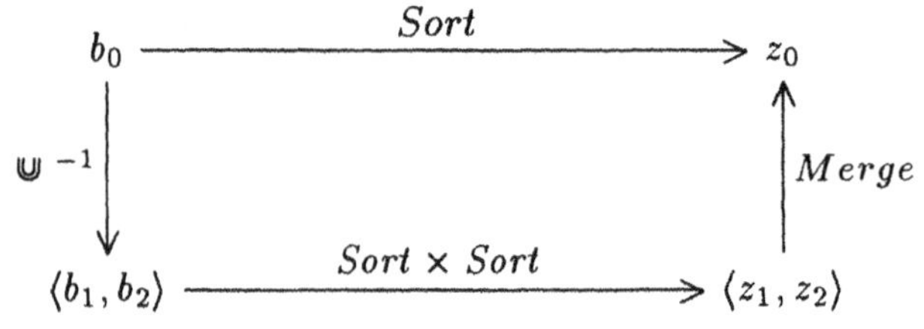


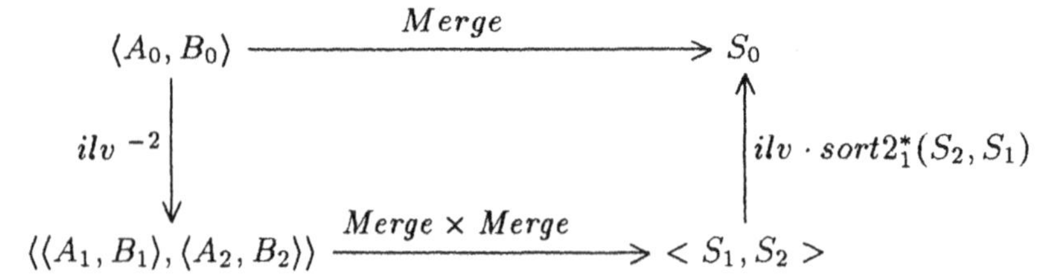

To simplify the analysis, assume that the input to *Sort* has length $n = 2^m$. Given $n$ processors, *Merge* runs in time

$$\begin{aligned} T_{Merge}(n) &= max(T_{Merge}(n/2), T_{Merge}(n/2)) + O(1) \\ &= O(log(n)) \end{aligned}$$

since the decomposition and composition operators both can be evaluated in constant time and the recursion goes to depth $O(log(n))$.

The decomposition operator $\uplus^{-1}$ in *Sort* is nondeterministic. This is an advantage at this stage of design since it allows us to defer committments and make choices that will maximize performance. In this case the complexity of *Sort* is calculated via the recurrence

$$T_{Sort}(n) = max(T_{Sort}(a(n)), T_{Sort}(b(n))) + O(log(n))$$

which is optimized by taking $a(n) = b(n) = n/2$ – that is, we split the input bag in half. Given $n$ processors this algorithm runs in $O(log^2(n))$ time, so it is suboptimal for sorting. However, according to [7], Batcher's Odd-Even sort is the most commonly used of parallel sort algorithms.

# 5 Related Sorting Algorithms

Several other parallel sorting algorithms can be developed using the techniques above. Batcher's bitonic sort [2] and the Periodic Balanced Sort [4] are also basically mergesort algorithms. They differ from Odd-Even sort in that the merge operation is a divide-and-conquer based on concatenation as the composition operator. For example, bitonic merge can be diagrammed as follows:

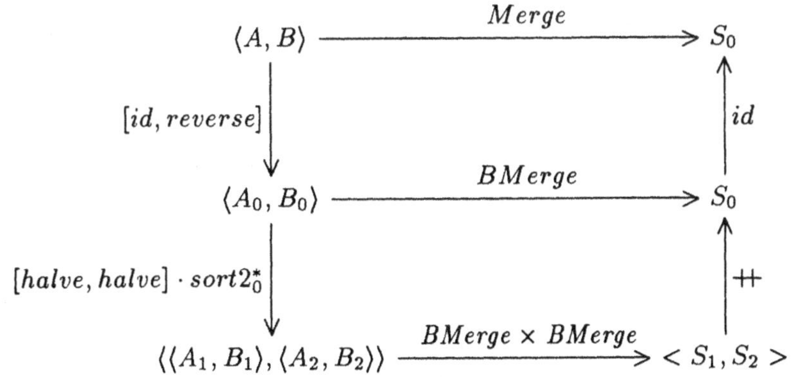


The essential fact about using ++ as a composition operator is that $\langle\langle A_1, B_1\rangle, \langle A_2, B_2\rangle\rangle$ must be a partition in the sense that no element of $A_1$ or $B_1$ is greater than any element of $A_2$ and $B_2$. The cleverness of the algorithm lies in a special property of sequences that allows a simple operation ($sort2_0^*$ here) to effectively produce a partition. This property is called "bitonicity" for bitonic sort and "balanced" for the periodic balanced sort. (The operation $\langle A_0, B_0\rangle = \langle id(A), reverse(B)\rangle$ establishes the bitonic property and decomposition preserves it). The challenge in deriving these algorithms lies in discovering these properties given that one wants a divide-and-conquer algorithm based on ++ as composition. Is there a systematic way to discover these properties or must we rely on creative invention? Admittedly, there may be other frameworks within which the discovery of these properties is easier.

Another well-known parallel sort algorithm is odd-even transposition sort. This can be viewed as a parallel variant of bubble-sort which in turn is derivable as a selection sort (local search is used to derive the selection subalgorithm). See the paper by Partsch in this volume.

The *ilv* constructor for sequences has many other applications including polynomial evaluation, discrete fast fourier transform, and matrix transposition. Butterfly and shuffle networks are natural architectures for implementing algorithms based on *ilv* [6].

# 6 Concluding Remarks

The Odd-Even sort algorithm is simpler to state than to derive. The properties of a *ilv*-based theory of sequences are much harder to understand and develop than a concatenation-based theory. However, the payoff is an abundance of algorithms with good parallel properties.

The derivation presented here requires a closer, more intensive development of the domain theory than most published derivations in the literature. The development was guided by some higher-level principles – invariance properties, distributive laws, and monotonicity laws provide most of the inference rules needed to support algorithm design.

We have used KIDS to derive a variant of the usual sequential mergesort algorithm [8]. However, simplifying assumptions in the implemented design tactic for divide-and-conquer disallow the derivation of the *ilv*-based merge described in Section 4. We are currently implementing a new algorithm design system based on [12] which overcomes these (and other) limitations and we see no essential difficulty in deriving the Odd-Even sort once the domain theory is place. Support for developing domain theories has not yet received enough serious attention in KIDS. For some theories, we have used KIDS to derive almost all of the laws needed to support the algorithm design process; other theories have been developed entirely by hand. In the current system, the theory development presented in Sections 3.1 and 4 would be done mostly manually.

The general message of this paper is that good parallel algorithms can be formally derived and that such derivations depend on the systematic development of the theory underlying the application domain. Furthermore, machine support can envisioned for both the theory development and algorithm derivation processes and this kind of support can be partially demonstrated at present.

Key elements of theory development are (1) defining basic concepts, operations, relations and the laws (axioms) that constrain their interpretation, (2) developing derived concepts, operations, and relations and important laws governing their behavior. The principle of seeking properties that are invariant under change or, conversely, operations that preserve important properties, provides strong guidance in theory development. In particular, distributive, monotonicity, and fixpoint laws are especially valuable and machine support for their acquisition is an important research topic.

# References


[1] AKL, S. *The Design and Analysis of Parallel Algorithms.* Prentice-Hall Inc., Englewood Cliffs, NJ, 1989.

[2] BATCHER, K. Sorting networks and their applications. In *AFIPS Spring Joint Computing Conference* (1968), vol. 32, pp. 307–314.

[3] BLAINE, L., AND GOLDBERG, A. DTRE – a semi-automatic transformation system. In *Constructing Programs from Specifications*, B. Möller, Ed. North-Holland, Amsterdam, 1991, pp. 165–204.

[4] DOWD, M., PERL, Y., RUDOLPH, L., AND SAKS, M. The periodic balanced sorting network. *Journal of the ACM 36*, 4 (October 1989), 738–757.

[5] GIBBONS, A., AND RYTTER, W. *Efficient Parallel Algorithms.* Cambridge University Press, Cambridge, 1988.

[6] JONES, G., AND SHEERAN, M. Collecting butterflies. Tech. Rep. PRG-91, Oxford University, Programming Research Group, February 1991.

[7] LEIGHTON, F. *Introduction to Parallel Algorithms and Architectures: Arrays, Trees, Hypercubes.* Morgan Kaufmann, San Mateo, CA, 1990.

[8] SMITH, D. R. Top-down synthesis of divide-and-conquer algorithms. *Artificial Intelligence 27*, 1 (September 1985), 43–96. (Reprinted in *Readings in Artificial Intelligence and Software Engineering*, C. Rich and R. Waters, Eds., Los Altos, CA, Morgan Kaufmann, 1986.).

[9] SMITH, D. R. KIDS – a semi-automatic program development system. *IEEE Transactions on Software Engineering Special Issue on Formal Methods in Software Engineering 16*, 9 (September 1990), 1024–1043.

[10] SMITH, D. R., AND LOWRY, M. R. Algorithm theories and design tactics. *Science of Computer Programming 14*, 2-3 (October 1990), 305–321.

[11] SMITH, D. R. Structure and design of problem reduction generators. In *Constructing Programs from Specifications*, B. Möller, Ed. North-Holland, Amsterdam, 1991, pp. 91–124.

[12] SMITH, D. R. Constructing specification morphisms. Tech. Rep. KES.U.92.1, Kestrel Institute, January 1992. to appear in Journal of Symbolic Computation, 1993.

# 3

# Some Experiments in Transforming Towards Parallel Executability

H.A. Partsch
University of Nijmegen
Department of Computing Science
Toernooiveld 1
NL-6525 ED Nijmegen
The Netherlands
e-mail: helmut@cs.kun.nl

**Abstract**

"Transformational Programming" summarizes a methodology for constructing correct and efficient programs from formal specifications by successive application of (provably) meaning-preserving transformation rules. This paper reports on several experiments on using this methodology, which has previously proven to be applicable to the development of sequential algorithms, for the formal derivation of algorithms executable on parallel architectures, notably SIMD machines. Based on the hypothesis that existing transformational knowledge should be sufficient also for deriving parallel algorithms, identifying useful transformation rules and strategic aspects of their use was one of the main goals of these experiments.

## 1 Introduction

"Transformational Programming" summarizes a methodology for constructing correct and efficient programs from formal specifications by successive application of (provably) meaning-preserving transformation rules. For an introduction and overview, cf., e.g., [5] or [8].

In the past, the methodology has proven to be successfully applicable to the systematic development of highly complicated sequential algorithms. Also, lots of claims have been made that transformational programming not only helps in developing correct sequential programs, but also profitably can be used to develop algorithms for various parallel architectures. Except for a few rather ad hoc experiments, however, almost none of these claims have been substantially and systematically backed by appropriate case studies. Our paper is primarily intended to provide such a case study.

The basic idea we are building on is outlined in [4]. It is based on the paradigm of functional programming and emphasizes the derivation of programs executable on a variety of different parallel architectures by employing a fixed repertoire of parallel program forms, called "skeletons". These skeletons, expressed as higher-order functions, play a dual role: they serve as a kind of "intermediate language" and thus provide a target for transforming specifications, and they can be seen as functional abstractions of underlying machine architectural features, i.e., certain classes of these functional forms characterize particular (classes of) parallel machines.
The idea is extremely simple and maybe even obvious. Nevertheless, it has not yet been exploited, although its advantages are at hand:

- The development of parallel algorithms can be divided into two well-defined and largely independent parts, viz. the derivation of programs formulated in terms of these functional forms from original problem specifications, and the efficient, maybe pre-defined implementation of these skeletons for concrete hardware. In this way, high-level aspects of algorithm design are disentangled from detailed architectural considerations. Thus, modifications of the original problem statement will only effect the first part of the development, whereas available implementations of skeletons are obviously open for reuse.
- It is possible to transform between different classes of these functional forms, thus providing a means to tackle the yet unsolved problem of portability between various substantially different parallel architectures.

In addition, for these activities existing transformational technology can be employed in order to assure correctness of the resulting "parallel program", which is to be seen as the ultimate and most important goal.

In our paper we report on experiments in applying the approach outlined above to the systematic derivation of certain parallel versions for a number of examples. In order to keep this case study at a reasonable length, we restrict ourselves to a particular form of parallel architectures, viz. the SIMD ("Single Instruction Multiple Data") machines, or, more accurately, the SFMD ("Single Function Multiple Data") paradigm (cf. [9]). The main purpose of our paper is in identifying interesting transformations to be used when transforming towards skeletons with an emphasis on the basic steps. Gaining elementary insight into strategic aspects may be seen as a secondary goal. Moreover, we start from the hypothesis that existing transformational knowledge should be sufficient to allow also the derivation of programs executable on parallel architectures. In this respect we go a step further than others (e.g. [9]) by trying to use not only the same methods, but also the same transformation rules – as far as possible.

There is a wide-spread misunderstanding about the benefits of such a kind of research which should be removed beforehand. The purpose of the paper is *not* inventing new algorithms or developing particularly clever ones, but rather trying to explore what kind of methodological knowledge is needed in order to formally derive known algorithms – although sometimes algorithms may come out of derivations which differ from those published in the literature (cf. final remark in section 4.4). Of course, as a long-term goal, it is intended to have available a kind of "tool-box" of transformations and strategies to be used for the derivation of unknown algorithms.

# 2 Preliminaries

As a basis for our following considerations we introduce some conventions on the notation of specifications and programs, as well as a few fundamental transformation rules. For brevity, we confine ourselves to a few important aspects. For a more comprehensive treatment and for details we refer the reader to [8].

## 2.1 Data types and expressions

For denoting specifications and programs, an expression language with strong typing is used. Certain basic *data types* (e.g., for specifying natural numbers, etc.) are assumed available and defined elsewhere (cf., e.g., again [8]). These data types specify object classes (such as **nat** for natural numbers) together with their characteristic operations the semantics of which is defined by algebraic axioms.

In addition to basic types, also basic *type schemes* will be used. These type schemes define structural characteristics of composed data structures (e.g., sets, sequences, etc.) independent of the types of their constituents. An example is the type scheme ESEQU ("extended sequences", cf. [8]) specifying an object class **sequ**, a constant <> (the empty sequence), and the operations (where dots indicate number and position of operands):

- .=<> test on emptyness;
- .≠<> test on non-emptyness;
- **first**. first element of a sequences;
- **rest**. sequence without its first element;
- .[.] indexed access;
- .[.:.] subsequence, determined by pair of indices (starting from 1);
- |.| length.

Additional operations on sequences are:

(invisible) "lifting" of elements (i.e., making elements into singleton sequences);
and
concatenation (denoted by juxtaposition).

Well-formed terms over constants, variables and operation symbols of basic types and type schemes are basic *expressions*. For forming more complex expressions additional well-known constructs such as

- conditional and guarded expression;
- function application;
- declaration of functions and objects; and
- abstraction

are used.

*Conditionals* and *guarded expressions* are denoted as usual. Within boolean expressions, occasionally $\Delta$ and $\nabla$, denoting "sequential and" and "sequential or", respectively, are used.

*Function applications* will also be written in a conventional style (using parentheses), in order to avoid the necessity of priority rules. An exception is formed by some operation symbols from the basic data types for which the usual priority rules are assumed.

*Function definitions* are denoted by giving the signature of the function (including name or operation symbol and functionality) and a defining equation. Thus, e.g., the length of a sequence is defined by a function

$|.|$: **sequ** $\rightarrow$ **nat**,
$|s| =_{\text{def}}$ **if** $s$ =<> **then** 0 **else** 1 + |**rest**$s$| **fi**.

Here $|.|$ is an operation symbol (with the dot indicating the position of the argument) for an operation that maps sequences to natural numbers.

Occasionally, definedness properties ("assertions") are stated to characterize partial functions. For the (partial) indexing operation $s[i]$, e.g., we have to require

**def**$(s[i]) \Rightarrow 1 \leq i \leq |s|$

in order to guarantee the definedness of an indexed access to a sequence.

Functions are also allowed to yield tuples as results. In order to be able to access the individual components of such a tuple result, an intuitively obvious selector notation (consisting of a dot and the number of the component to be selected) will be used.

In order to be able to formulate higher-order functions we need *functional abstractions*. These are denoted by expressions preceded by their functionality. A simple example is the higher-order function

*str*: **sequ** $\rightarrow$ (**nat** $\rightarrow$ **char**),
**def**$(str(s)) \Rightarrow 1 \leq i \leq |s|$,
$str(s) =_{\text{def}}$ (**nat** $i$) **char**: $s[i]$

which maps a sequence to a (partial) function that yields for an (admissible) index the respective character of the string.

## 2.2 Basic transformations

In order to be able to perform transformational developments we need basic transformation rules. A comprehensive collection of such rules can be found in [8]. In the following we give a few examples all of which will be used later on.

The most elementary transformation rules are "unfold" and "fold" (cf. [3]). In our setting they are formulated as follows:

*Unfold* (for functions)

$f$(E)

↓ [ DEF[E], DET[E]

E'[E **for** $x$]

**Syntactic constraints:**
NOTOCCURS[$x$ **in** E],
DECL[$f$] = $f$: **m** → **n**, $f(x) =_{def}$ E'

*Fold* (for functions)

E'[E **for** $x$]

↓ [ DEF[E]

$f$(E)

**Syntactic constraints:**
NOTOCCURS[$x$ **in** E],
DECL[$f$] = $f$: **m** → **n**, $f(x) =_{def}$ E'

As can be seen from these two examples, transformation rules are denoted by two (schematic) expressions, the "input scheme" (above the arrow) and the "output scheme" (below the arrow), respectively, and an arrow which characterizes the semantic relationship (between these expressions) that is established by the rule. A bidirectional arrow (cf. below) denotes semantic equivalence, a unidirectional one descendance (cf. [8]), a kind of refinement. The output scheme is followed by a (possibly empty) list of syntactic constraints, such as:

- NOTOCCURS, an identifier does not occur in an expression; or
- DECL, yielding the declaration for an identifier.

To the right of the arrows, semantic applicability conditions are mentioned, e.g. above:

- DEF, an expression has a defined value;
- DET, an expression is determinate.

For both, syntactic constraints as well as semantic applicability conditions, the availability of predefined predicate symbols (such as DEF, DET, etc.) on expressions is assumed. Furthermore, it is assumed that all free variables within these conditions are universally quantified.

In order to avoid notational overhead through trivial applicability conditions, for all constituents of transformation rules (i.e., input scheme, output scheme, and applicability conditions), syntactic correctness and context-correctness are assumed as a general convention.

Similarly to the rules above, other ones can be given, e.g.:

- rules on conditionals, e.g.

<u>*Simplification of a conditional*</u>

**if** B **then** $E_1$ **else** $E_2$ **fi**

$\Updownarrow$ [ B ≡ **true**

$E_1$

- distributivity rules, such as

<u>*Distributivity of function call over conditional*</u>

*f*(**if** B **then** $E_1$ **else** $E_2$ **fi**)

$\Updownarrow$

**if** B **then** *f*($E_1$) **else** *f*($E_2$) **fi**

- embedding ("generalization")

<u>*Generalization*</u>

*f*(E*) **where**
*f*: **m** → **n**,
$f(x) =_{def} E'(x)$

$\Updownarrow$ [ $E''(x, E) \equiv E'(x)$

*f'*(E*, E) **where**
*f'*:(**m** × **p**) → **n**,
$f'(x, y) =_{def} E''(x, y)$

*Totalization of a partial function*

$f$(E) **where**
$f$: **m** $\rightarrow$ **n**, **def**($f(x)$) $\Rightarrow$ P($x$),
$f(x) =_{def}$ E'

$\Updownarrow$ [ DEF[E], DET[E], P(E) $\equiv$ **true**

$g$(E) **where**
$g$: **m** $\rightarrow$ **n**,
$g(x) =_{def}$ **if** P($x$) **then** $f(x)$ **else** E" **fi**

– currying (on the first argument, with analogous rules for other arguments)

*Currying*

$f$(E', E") **where**
$f$: (**m** $\times$ **m'**) $\rightarrow$ **n**, $f(x, y) =_{def}$ E

$\Updownarrow$

$f'$(E')(E") **where**
$f'$: **m** $\rightarrow$ (**m'** $\rightarrow$ **n**), $f'(x) =_{def}$ ((**m'** $y$) **n**: E[$f'$(A)(B) **for** $f$(A, B)])

*Currying* (variant for functions defined by conditional)

$f$(E', E") **where**
$f$: (**m** $\times$ **m'**) $\rightarrow$ **n**, $f(x, y) =_{def}$ **if** T($x$) **then** $E_1$ **else** $E_2$ **fi**

$\Updownarrow$

$f'$(E')(E") **where**
$f'$: **m** $\rightarrow$ (**m'** $\rightarrow$ **n**),
$f'(x) =_{def}$ **if** T($x$) **then** ((**m'** $y$) **n**: $E_1$[$f'$(A)(B) **for** $f$(A, B)])
**else** ((**m'** $y$) **n**: $E_2$[$f'$(A)(B) **for** $f$(A, B)]) **fi**

- recursion simplification

<u>*Inversion*</u>

$$f\colon \mathbf{m} \rightarrow \mathbf{n},$$
$$f(x) =_{\mathrm{def}} \mathbf{if}\ x{=}\mathrm{E}\ \mathbf{then}\ \mathrm{H}(x)\ \mathbf{else}\ p(f(\mathrm{K}(x)))\ \mathbf{fi}$$

$$\Updownarrow \quad \left[\ \begin{array}{l} \mathsf{DEF}[\mathrm{E}],\ \mathsf{DET}[\mathrm{E}], \\ \mathsf{DEF}[\mathrm{K}(x)]\ \vdash\ (\mathrm{K}^{-1}(\mathrm{K}(x)) = x) \equiv \mathbf{true} \end{array} \right.$$

$$f\colon \mathbf{m} \rightarrow \mathbf{n},$$
$$f(x) =_{\mathrm{def}} g(x, \mathrm{E}, \mathrm{H}(\mathrm{E}))\ \mathbf{where}$$
$$\quad g\colon (\mathbf{m} \times \mathbf{m} \times \mathbf{n}) \rightarrow \mathbf{n}:$$
$$\quad g(x, y, z) =_{\mathrm{def}} \mathbf{if}\ y{=}x\ \mathbf{then}\ z\ \mathbf{else}\ g(x, \mathrm{K}^{-1}(y), p(z))\ \mathbf{fi}$$

- simplification (according to domain knowledge), e.g.

  $|<>| \equiv 0$

  or by combining several elementary steps.

In all these rules **m**, **n**, etc., denote arbitrary types which, in particular, also can be tuples or functional types. Thus, in particular, most of the rules above also can be applied to higher-order functions.

## 2.3 Strategic considerations

The global strategy to be used within our sample developments is mainly determined by the goal we want to reach: functionals defined in terms of (pre-defined) skeletons using first-order function application, function composition, conditionals, tupling, abstraction, and **where**-abbreviations.

Individual steps within this general guideline depend on the form of the specification we start from. In case of an already functional specification, we aim at a transition to a composition of functional forms which can basically be achieved by folding with the available skeleton definitions. If the starting point of the development is an operational, applicative specification, a lifting to the functional level (essentially to be achieved by currying) is the most important activity. Frequently, this has to be preceded by some preparatory steps to remove dependencies betwen the parameters. For an initally given, non-operational specification, the development of an equivalent, applicative or functional specification has to be performed first.

Although finally aiming at automatic strategies as far as possible, we use a purely manual strategy in our sample develoments below, because not yet enough information on this important topic is available so far.

# 3 Some functional forms characterizing SFMD architectures

A systematic comparison of various parallel architectures and their representation by corresponding functional forms can be found in [4]. Since we are aiming at algorithms for SFMD architectures, we confine ourselves to functional forms typical for this kind of machines which are basically characterized by vectors (or arrays) and operations on these. However, rather than defining vectors by a suitable data structure, we prefer to represent them by functions on finite domains, which has the advantage that available transformations (for functionals) can be used, and a particular "problem theory" (cf. [11]) for arrays (comprising theorems to be used in a derivation) is not required.

In particular, let

$$\mathbf{inat} = (\mathbf{nat}\ n\text{: } l \leq n \leq h)$$

denote a (finite, non-empty, linearly ordered) index domain. Then functions $a$ of type

$$\mathbf{inat} \to \mathbf{m}$$

can be viewed as functional descriptions of vectors or arrays with index domain **inat**, i.e., visualized as

| | | | |
|---|---|---|---|
| $a_l$ | $a_{l+1}$ | ... | $a_h$ |

## 3.1 Basic skeletons

As outlined above, operations on vectors are described by functionals, called "skeletons". There are various basic skeletons to describe SIMD architectures, i.e., typical functionals on functions of type **inat** → **m**, which can be grouped according to their effect into different classes as given below.

For some of the definitions a visual aid (for **inat** with $l = 1$ and $h = n$) is given by sketching the essential effect of these functionals on the vectors involved (while disregarding other parameters of the respective functionals).

*3.1.1. Functionals resulting in elementary ("scalar") values*

*Lower bound:*

$L$: (**inat** → **m**) → **inat**,
$L(a) =_{\text{def}} l$,

*Higher bound:*

$H$: (**inat** → **m**) → **inat**,
$H(a) =_{\text{def}} h$,

*Selection of a component:*

$SEL$: ((**inat** $\rightarrow$ **m**) $\times$ **inat**) $\rightarrow$ **m**,
$SEL(a, i) =_{def} a(i)$,

*Length of a vector:*

$LEN$: (**inat** $\rightarrow$ **m**) $\rightarrow$ **nat**,
$LEN(a) =_{def} H(a)-L(a)+1$

*3.1.2. Functionals that retain unchanged the size of their domains (independent of their element values)*

*Creation of a new constant vector:*

$NEW$: **m** $\rightarrow$ (**inat** $\rightarrow$ **m**),
$NEW(v) =_{def}$ (**inat** $i$) **m**: $v$,

*Creation of an "identity" vector:*

$INEW$: $\rightarrow$ (**inat** $\rightarrow$ **inat**),
$INEW =_{def}$ (**inat** $i$) **inat**: $i$,

*Application of an operation to all elements of a vector:*

$MAP$: ((**m** $\rightarrow$ **n**) $\times$ (**inat** $\rightarrow$ **m**)) $\rightarrow$ (**inat** $\rightarrow$ **n**),
$MAP(f, a) =_{def}$ (**inat** $i$) **n**: $f(a(i))$,

*Linear shift of the elements of a vector:*

$SHIFTL$: ((**inat** $\rightarrow$ **m**) $\times$ **nat** $\times$ **m**) $\rightarrow$ (**inat** $\rightarrow$ **m**),
**def**$(SHIFTL(a, d, v)) \Rightarrow d \leq LEN(a)$,
$SHIFTL(a, d, v) =_{def}$ (**inat** $i$) **m**: **if** $i > H(a)-d$ **then** $v$ **else** $a(i+d)$ **fi**,

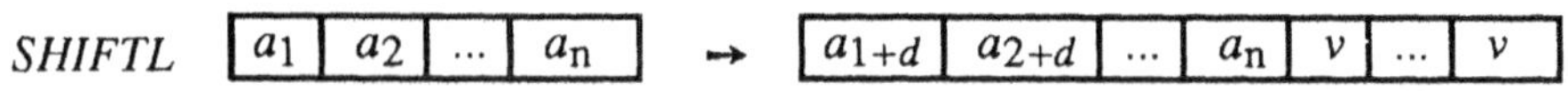


$SHIFTR$: ((**inat** $\rightarrow$ **m**) $\times$ **nat** $\times$ **m**) $\rightarrow$ (**inat** $\rightarrow$ **m**),
**def**$(SHIFTR(a, d, v)) \Rightarrow d \leq LEN(a)$,
$SHIFTR(a, d, v) =_{def}$ (**inat** $i$) **m**: **if** $i < L(a)+d$ **then** $v$ **else** $a(i-d)$ **fi**,

*Cyclic shift of the elements of a vector:*

*CSHIFTL*: ((**inat** → **m**) × **nat**) → (**inat** → **m**),
*CSHIFTL*$(a, d) =_{\text{def}}$ (**inat** $i$) **m**: $a(((i–L(a)+d)$ **mod** $LEN(a))+L(a))$,

*CSHIFTL* 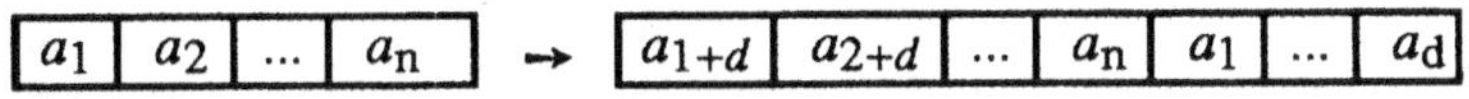


*CSHIFTR*: ((**inat** → **m**) × **nat**) → (**inat** → **m**),
*CSHIFTR*$(a, d) =_{\text{def}}$ (**inat** $i$) **m**: $a(((i–L(a)+LEN(a)–d)$ **mod** $LEN(a))+L(a))$,

*"Pairing" and "unpairing" of vectors:*

*ZIP*: ((**inat** → **m**) × (**inat** → **n**)) → (**inat** → (**m** × **n**)),
*ZIP*$(a, b) =_{\text{def}}$ (**inat** $i$) (**m** × **n**): $(a(i), b(i))$,

*ZIP* 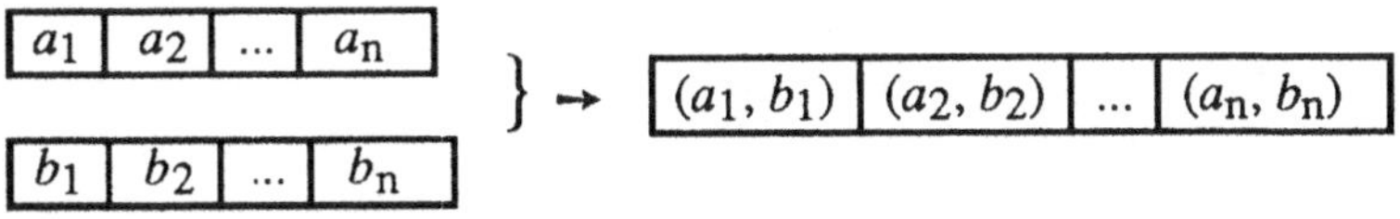


*UNZIP*: (**inat** → (**m** × **n**)) → ((**inat** → **m**) × (**inat** → **n**)),
*UNZIP*$(c) =_{\text{def}}$ ((**inat** $i$) **m**: $c(i).1$, (**inat** $i$) **n**: $c(i).2$),

*UNZIP* 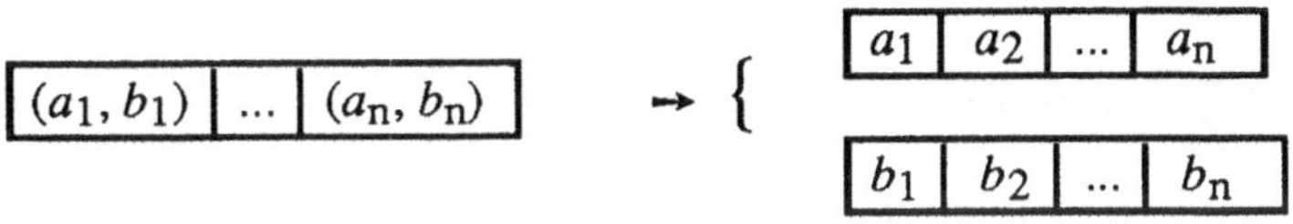


*3.1.3. Functionals that change the size of their domains (independent of the values of the elements)*

*Extending a vector:*

*EXT*: ((**inat** → **m**) × **nat** × **m**) → (**iinat** → **m**),
*EXT*$(a, d, v) =_{\text{def}}$ (**iinat** $i$) **m**: **if** $i > H(a)$ **then** $v$ **else** $a(i)$ **fi where**
**iinat** = (**nat** $n$: $L(a) \leq n \leq H(a)+d$)

*EXT* 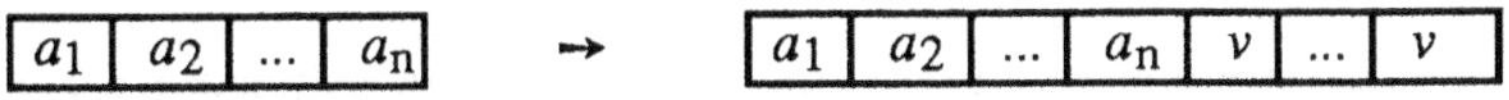


*Truncating a vector:*

*TRUNC*: ((**inat** → **m**) × **nat**) → (**iinat** → **m**),
**def**(*TRUNC*$(a, d)) \Rightarrow d \leq LEN(a)$,
*TRUNC*$(a, d) =_{\text{def}}$ (**iinat** $i$) **m**: $a(i)$ **where**
**iinat** = (**nat** $n$: $L(a) \leq n \leq H(a)–d$)

*TRUNC* 

*3.1.4. Functionals that change the size of their domains (dependent on the values of the elements)*

In [4] several skeletons that change the domain of a vector dependent on the values of its components are given. Examples are the skeletons *FETCH* and *SEND* for extracting, resp. updating, a certain subvector. Since these functionals will not be needed in our sample developments, we skip their formal definitions.

*3.1.5. "Test"-functionals*

These skeletons will be used as shorthand notation to formulate definedness conditions in connection with partiality. Examples are

*A vector represents an injective mapping:*

$INJ$: $(\textbf{inat} \to \textbf{m}) \to \textbf{bool}$,
$INJ(a) =_{\text{def}} \forall\ \textbf{inat}\ i, j: a(i) = a(j) \Rightarrow i = j$,

*All components of a vector are the same:*

$CONST$: $(\textbf{inat} \to \textbf{m}) \to \textbf{bool}$,
$CONST(a) =_{\text{def}} \forall\ \textbf{inat}\ i, j: a(i) = a(j)$.

## 3.2 Derived skeletons

The skeleton definitions given in the previous section not only characterize certain machine primitives, but also serve as a basis for defining additional functionals which also may profitably be used within concrete developments. Below we give a few examples of this kind. For all of them we first give an independent definition of such additional functionals and then show how definitions in terms of the basic skeletons can be derived by simple application of transformations. Thus, as a byproduct, transformational methodology is exemplified. Further examples of the same kind can be found in [4].

*3.2.1 Variants of the* MAP *skeleton*

In the previous section a simple definition of a functional *MAP* has been given where a function is applied to all components of a vector. Using this definition as a basis, a lot of useful variants (expressed in terms of *MAP*) can be derived which differ in the kind of base function that is to be applied to all components.

Application of a base function that also depends on the index domain can be specified by

$IMAP$: $(((\mathbf{inat} \times \mathbf{m}) \to \mathbf{n}) \times (\mathbf{inat} \to \mathbf{m})) \to (\mathbf{inat} \to \mathbf{n})$,
$IMAP(f, a) =_{\text{def}} (\mathbf{inat}\ i)\ \mathbf{n}{:}\ f(i, a(i))$.

A definition in terms of basic skeletons is derived as follows (where a hint on the transformation used is given between "[" and "]" following the "≡" symbol):

$(\mathbf{inat}\ i)\ \mathbf{n}{:}\ f(i, a(i))$

$\equiv$ [ fold *MAP* ]

$MAP(f, (\mathbf{inat}\ i)\ (\mathbf{inat} \times \mathbf{m}){:}\ (i, a(i)))$

$\equiv$ [ fold *ZIP* ]

$MAP(f, ZIP((\mathbf{inat}\ i)\ \mathbf{inat}{:}\ i, a))$

$\equiv$ [ fold *INEW* ]

$MAP(f, ZIP(INEW, a))$.

Application of a base function with two arguments can be specified by

$MAP2$: $(((\mathbf{m} \times \mathbf{n}) \to \mathbf{r}) \times (\mathbf{inat} \to \mathbf{m}) \times (\mathbf{inat} \to \mathbf{n})) \to (\mathbf{inat} \to \mathbf{r})$,
$MAP2(f, a, b) =_{\text{def}} (\mathbf{inat}\ i)\ \mathbf{r}{:}\ f(a(i), b(i))$.

A definition in terms of basic skeletons is derived as follows:

$(\mathbf{inat}\ i)\ \mathbf{r}{:}\ f(a(i), b(i))$

$\equiv$ [ fold *MAP* ]

$MAP(f, (\mathbf{inat}\ i)(\mathbf{m} \times \mathbf{n}){:}\ (a(i), b(i)))$

$\equiv$ [ fold *ZIP* ]

$MAP(f, ZIP(a, b))$.

Application of a base function with two arguments and two results can be specified by

$MAP2\text{-}2$: $(((\mathbf{m} \times \mathbf{n}) \to (\mathbf{r} \times \mathbf{p})) \times (\mathbf{inat} \to \mathbf{m}) \times (\mathbf{inat} \to \mathbf{n})) \to$
$((\mathbf{inat} \to \mathbf{r}) \times (\mathbf{inat} \to \mathbf{p}))$,
$MAP2\text{-}2(f, a, b) =_{\text{def}} ((\mathbf{inat}\ i)\ \mathbf{r}{:}\ (f(a(i), b(i))).1, (\mathbf{inat}\ i)\ \mathbf{p}{:}\ (f(a(i), b(i))).2)$.

A definition in terms of basic skeletons is derived as follows:

$((\mathbf{inat}\ i)\ \mathbf{r}{:}\ (f(a(i), b(i))).1, (\mathbf{inat}\ i)\ \mathbf{p}{:}\ (f(a(i), b(i))).2)$

$\equiv$ [ fold *UNZIP* ]

$UNZIP((\mathbf{inat}\ i)(\mathbf{r} \times \mathbf{p}){:}\ f(a(i), b(i)))$

$\equiv$ [ fold *MAP* ]

$UNZIP(MAP(f, (\mathbf{inat}\ i)(\mathbf{r} \times \mathbf{p}){:}\ (a(i), b(i))))$

$\equiv$ [ fold *ZIP* ]

$UNZIP(MAP(f, ZIP(a, b)))$

*3.2.2. Variants of the (cyclic)* SHIFT *functionals*

Different from the linear shift operations where, for obvious reasons, it is required that the "length of the shift" is at most the length of the array, the cyclic shift operations do not need such a restriction. If, however, we demand the same restriction, different definitions of the cyclic shift operations can be derived from the old definitions.

For the particular case

$$a < 2b$$

the usual definition of **mod**,

$$a \ \mathbf{mod}\ b =_{\text{def}} \mathbf{if}\ a < b\ \mathbf{then}\ a\ \mathbf{else}\ (a-b)\ \mathbf{mod}\ b\ \mathbf{fi},$$

simplifies to

$$a \ \mathbf{mod}\ b =_{\text{def}} \mathbf{if}\ a < b\ \mathbf{then}\ a\ \mathbf{else}\ a-b\ \mathbf{fi}$$

which can be used to derive new definitions of the cyclic shift operations from the definitions given in the previous section:

$CSHIFTL(a, d)$

≡ [ definition of *CSHIFTL* ]

$(\mathbf{inat}\ i)\ \mathbf{m}\text{: } a(((i-L(a)+d)\ \mathbf{mod}\ LEN(a))+L(a))$

≡ [ $d \le LEN(a) \wedge 0 \le i-L(a) < LEN(a) \Rightarrow i-L(a)+d < 2LEN(a)$;
simplified definition of **mod** ]

$(\mathbf{inat}\ i)\ \mathbf{m}\text{: } a(\mathbf{if}\ i-L(a)+d < LEN(a)\ \ \mathbf{then}\ i-L(a)+d$
$\mathbf{else}\ i-L(a)+d-LEN(a)\ \mathbf{fi} + L(a))$

≡ [ distributivity of operation and function call over conditional ]

$(\mathbf{inat}\ i)\ \mathbf{m}\text{: } \mathbf{if}\ i-L(a)+d < LEN(a)\ \ \mathbf{then}\ a(i-L(a)+d+L(a))$
$\mathbf{else}\ a(i-L(a)+d-LEN(a)+L(a))\ \mathbf{fi}$

≡ [ simplification (using definition of *LEN*) ]

$(\mathbf{inat}\ i)\ \mathbf{m}\text{: } \mathbf{if}\ i+d \le H(a)\ \mathbf{then}\ a(i+d)\ \mathbf{else}\ a(i+d-LEN(a))\ \mathbf{fi}.$

Thus, as an alternative (partial) definition for *CSHIFTL* we have

$CSHIFTL\text{: } ((\mathbf{inat} \to \mathbf{m}) \times \mathbf{nat}) \to (\mathbf{inat} \to \mathbf{m}),$
$\mathbf{def}(CSHIFTL(a, d)) \Rightarrow d \le LEN(a)),$
$CSHIFTL(a, d) =_{\text{def}} (\mathbf{inat}\ i)\ \mathbf{m}\text{: } \mathbf{if}\ i+d \le H(a)\ \ \mathbf{then}\ a(i+d)$
$\mathbf{else}\ a(i+d-LEN(a))\ \mathbf{fi}.$

Analogously, an alternative (partial) definition for *CSHIFTR* can be derived:

$CSHIFTR$: ((**inat** → **m**) × **nat**) → (**inat** → **m**),
**def**($CSHIFTR(a, d,)$) ⇒ $d \le LEN(a)$),
$CSHIFTR(a, d) =_{def}$ (**inat** $i$) **m**: **if** $i \ge L(a)+d$ **then** $a(i-d)$
**else** $a(i-d+LEN(a))$ **fi**.

*3.2.3. Updating a vector*

A last example of derived functionals is provided by an operation to update an element of a vector:

$UPDATE$: ((**inat** → **m**) × **inat** × **m**) → (**inat** → **m**),
$UPDATE(a, k, v) =_{def}$ (**inat** $i$) **m**: **if** $i = k$ **then** $v$ **else** $a(i)$ **fi**

which can be transformed (by folding $IMAP$) into

$UPDATE(a, k, v) =_{def} IMAP$((**inat** $i$, **m** $x$) **m**: **if** $i = k$ **then** $v$ **else** $x$ **fi**, $a$).

# 4 Transforming specifications into functional forms

In this section we deal with a collection of examples, mainly in order to exemplify how program transformations (as given in section 2.2) can be used to derive programs expressed in terms of skeletons (as defined in section 3.1). In particular, we will demonstrate how the development process is influenced by the form of the specification we start from. Therefore, the examples given below are to be considered as representatives of classes of specifications such as functional, applicative, or non-operational descriptive specifications.

## 4.1 A simple problem

As a first problem we consider a simple, nearly artificial example of a functional specification. It will turn out that straightforward transformation steps are sufficient to achieve significant improvement with respect to performance on parallel architectures.

Intuitively, the problem is to compute all values of a binary function $f$ applied to all pairs from the Cartesian product of two sets (of values of type **m** and **n**, respectivley) of the same (finite) cardinality $I+1$.

Formally the problem can be stated as follows:
Given

**inat** = (**nat** $i$: $0 \le i \le I$),
$a$: **inat** → **m**, injective,
$b$: **inat** → **n**, injective,
$f$: (**m** × **n**) → **r**,

compute

$$\{\mathbf{r}\ x\colon \exists\ \mathbf{inat}\ i, j\colon x = F(f, a, b)(i, j)\} \qquad (4.1\text{-}1)$$

where

$$F\colon (((\mathbf{m} \times \mathbf{n}) \to \mathbf{r}) \times (\mathbf{inat} \to \mathbf{m}) \times (\mathbf{inat} \to \mathbf{n})) \to ((\mathbf{inat} \times \mathbf{inat}) \to \mathbf{r}),$$
$$\mathbf{def}(F(f, g, g')) \Rightarrow INJ(g) \wedge INJ(g'),$$
$$F(f, g, g') =_{\mathrm{def}} (\mathbf{inat}\ i, \mathbf{inat}\ j)\ \mathbf{r}\colon f(g(i), g'(j)).$$

A direct conversion of $F$ into a composition of functional forms is possible by simple folding, viz.

$$F(f, g, g') =_{\mathrm{def}} (\mathbf{inat}\ i, \mathbf{inat}\ j)\ \mathbf{r}\colon f(SEL(g, i), SEL(g', j)).$$

The result, however, is obviously not well suited for parallel execution, since for computing (4.1-1) $F(f, g, g')$ has to be applied to all possible pairs of indices $i$ and $j$. Thus, we have a complexity of order $O(I^2)$, even in a parallel environment.

By currying, however, (4.1-1) can be transformed into

$$\{\mathbf{r}\ x\colon \exists\ \mathbf{inat}\ i, j\colon x = F'(f, a, b)(i)(j)\} \qquad (4.1\text{-}2)$$

where

$$F'\colon (((\mathbf{m} \times \mathbf{n}) \to \mathbf{r}) \times (\mathbf{inat} \to \mathbf{m}) \times (\mathbf{inat} \to \mathbf{n})) \to (\mathbf{inat} \to (\mathbf{inat} \to \mathbf{r})),$$
$$\mathbf{def}(F'(f, g, g')) \Rightarrow INJ(g) \wedge INJ(g'),$$
$$F'(f, g, g') =_{\mathrm{def}} (\mathbf{inat}\ i)\ (\mathbf{inat} \to \mathbf{r})\colon (\mathbf{inat}\ j)\ \mathbf{r}\colon f(g(i), g'(j)).$$

Now a conversion into a composition of functional forms is possible by a sequence of fold steps:

$$(\mathbf{inat}\ j)\ \mathbf{r}\colon f(g(i), g'(j))$$

$\equiv$ [ fold $MAP$ ]

$$MAP(f, (\mathbf{inat}\ j)\ (\mathbf{m} \times \mathbf{n})\colon (g(i), g'(j)))$$

$\equiv$ [ fold $ZIP$ ]

$$MAP(f, ZIP((\mathbf{inat}\ j)\ \mathbf{m}\colon g(i), g'))$$

$\equiv$ [ fold $NEW$ ]

$$MAP(f, ZIP(NEW(g(i)), g'))$$

$\equiv$ [ fold $SEL$ ]

$$MAP(f, ZIP(NEW(SEL(g, i)), g')).$$

Thus, we obtain

$$F'(f, g, g') =_{\mathrm{def}} (\mathbf{inat}\ i)\ (\mathbf{inat} \to \mathbf{r})\colon MAP(f, ZIP(NEW(SEL(g, i)), g')))$$

which obviously is well suited for parallel execution. Here (4.1-1) is obtained by applying $F'(f, g, g')$ to all indices $i$ which can be done with complexity of order $O(I)$ in a parallel environment, since the value in each column below are computed in parallel:

| | | | |
|---|---|---|---|
| $f(g(0), g'(0))$ | $f(g(1), g'(0))$ | | $f(g(I), g'(0))$ |
| $f(g(0), g'(1))$ | $f(g(1), g'(1))$ | | $f(g(I), g'(1))$ |
| $\vdots \; , \; \vdots$ | $\vdots \; , \; \vdots$ | ... | $\vdots \; , \; \vdots$ |
| $f(g(0), g'(I-1))$ | $f(g(1), g'(I-1))$ | | $f(g(I), g'(I-1))$ |
| $f(g(0), g'(I))$ | $f(g(1), g'(I))$ | | $f(g(I), g'(I))$ |

For the particular example we are dealing with in this section, a further reduction of the order of complexity is impossible. However, the number of basic skeletons in computing $F$ can be further reduced. To this end, we use the fact that (4.1-1) is also equivalent to

$$\{\mathbf{r}\ x: \exists\ \mathbf{inat}\ i, j: x = F''(f, a, b)(i)(j)\} \tag{4.1-3}$$

where

$F''$: $(((\mathbf{m} \times \mathbf{n}) \to \mathbf{r}) \times (\mathbf{inat} \to \mathbf{m}) \times (\mathbf{inat} \to \mathbf{n})) \to (\mathbf{inat} \to (\mathbf{inat} \to \mathbf{r}))$,
$\mathbf{def}(F'(f, g, g')) \Rightarrow INJ(g) \wedge INJ(g')$,
$F''(f, g, g') =_{\text{def}} (\mathbf{inat}\ i)\ (\mathbf{inat} \to \mathbf{r})$: $(\mathbf{inat}\ j)\ \mathbf{r}$: $f(g(j+i\ \mathbf{mod}\ (I+1)), g'(j))$.

Again, a conversion into a composition of functional forms by successive foldings is possible:

$(\mathbf{inat}\ j)\ \mathbf{r}$: $f(g(j+i\ \mathbf{mod}\ (I+1)), g'(j))$

$\equiv$ [ fold $MAP$ ]

$MAP(f, (\mathbf{inat}\ j)\ (\mathbf{m} \times \mathbf{n})$: $(g(j+i\ \mathbf{mod}\ (I+1)), g'(j)))$

$\equiv$ [ fold $ZIP$ ]

$MAP(f, ZIP((\mathbf{inat}\ j)\ \mathbf{m}$: $g(j+i\ \mathbf{mod}\ (I+1))), g'))$

$\equiv$ [ fold $CSHIFTL$ ]

$MAP(f, ZIP(CSHIFTL(g, i)), g'))$.

Thus, we obtain

$F''(f, g, g') =_{\text{def}} (\mathbf{inat}\ i)\ (\mathbf{inat} \to \mathbf{r})$: $MAP(f, ZIP(CSHIFTL(g, i)), g'))$

which is even better suited for parallel execution, since the number of skeletons used is less than before, and, in particular, no scalar operations are used. Now the computation of $F''$ proceeds as follows:

$$
\begin{array}{llllll}
f(g(0), & g'(0)) & f(g(1), & g'(0)) & & f(g(I), & g'(0)) \\
f(g(1), & g'(1)) & f(g(2), & g'(1)) & & f(g(0), & g'(1)) \\
\;\vdots\;, & \vdots & \;\vdots\;, & \vdots & \cdots & \;\vdots\;, & \vdots \\
f(g(I-1), & g'(I-1)) & f(g(I), & g'(I-1)) & & f(g(I-2), & g'(I-1)) \\
f(g(I), & g'(I)) & f(g(0), & g'(I)) & & f(g(I-1), & g'(I))
\end{array}
$$

Already in the context of this simple artifical problem, two important observations with respect to strategy can be made. In order to improve the performance of a given functional specification within a parallel environment

- transformation to functionals by currying (i.e., changing functionality); and
- internal transformations (i.e., keeping functionality but changing definitions)

are straightforward and often very effective techniques.

## 4.2 Binomial Coefficient

The second example we want to deal with, the well-known binomial coefficient, is intended to demonstrate how a given applicative specification can be transformed into a composition of skeletons. Apart from again using currying to "lift" the specification to the functional level, we also will see that often preparatory work is necessary to make dependent parameters of a function independent – which is a mandatory prerequisite for achieving parallel executability through the application of currying.

In the following, we assume natural numbers $I$ and $J$ such that $J \leq I$. An initial specification of our sample problem is then provided by

*bin*($I$, $J$) **where**

**inat** = (**nat** $i$: $0 \leq i \leq I$),

*bin*: (**inat** × **inat**) → **nat**,
**def**(*bin*($i$, $j$)) ⇒ $0 \leq j \leq i$,
*bin*($i$, $j$) $=_{\text{def}}$ **if** $i = 0 \vee j = 0 \vee j = i$ **then** 1 **else** *bin*($i-1$, $j-1$) + *bin*($i-1$, $j$) **fi**.

In our subsequent transformational derivation, we not only give the particular rules (referring to section 2.2 and [8]) and the results of their appliation, but also try to provide a kind of rationale of the development by explicitly stating various intermediate goals to be achieved. In a more abstract view, the collection of these intermediate goals is to be seen as a basis for a general strategy.

**general goal:** transformation of the original specification into an equivalent definition composed of functional forms

**1. subgoal:** transformation into an equivalent definition with independent parameters

➤ Totalization of a partial function

*bin'*(*I*, *J*) **where**

*bin'*: (**inat** × **inat**) → **nat**,
$bin'(i, j) =_{\text{def}}$ **if** $i \ge j$ **then** $bin(i, j)$ **else** 0 **fi**

➤ *Case introduction* in *bin'* ($i = 0 \lor i > 0$)

$bin'(i, j) =_{\text{def}}$ **if** $i = 0$ **then if** $i \ge j$ **then** $bin(i, j)$ **else** 0 **fi**
**else if** $i \ge j$ **then** $bin(i, j)$ **else** 0 **fi fi**

➤ *Simplification* in **then**-branch (premise: $i = 0$);
*Case introduction* in **else**-branch ($j = 0 \lor j > 0$)

$bin'(i, j) =_{\text{def}}$ **if** $i = 0$ **then if** $j = 0$ **then** 1 **else** 0 **fi**
**else if** $j = 0$ **then if** $i \ge j$ **then** $bin(i, j)$ **else** 0 **fi**
**else if** $i \ge j$ **then** $bin(i, j)$ **else** 0 **fi fi fi**

➤ *Transformation* of **else-then**-branch into expression in terms of *bin'* (under premise $i > 0 \land j = 0$):

**if** $i \ge j$ **then** $bin(i, j)$ **else** 0 **fi**

≡ [ $i \ge j \equiv i-1 \ge j$; $bin(i, j) \equiv 1 \equiv bin(i-1, j)$ ]

**if** $i-1 \ge j$ **then** $bin(i-1, j)$ **else** 0 **fi**

≡ [ fold *bin'* ]

$bin'(i-1, j)$

*Transformation* of **else-else**-branch into expression in terms of *bin'* (under premise $i > 0 \land j > 0$):

**if** $i \ge j$ **then** $bin(i, j)$ **else** 0 **fi**

≡ [ case introduction ]

**if** $i = j$ **then if** $i \ge j$ **then** $bin(i, j)$ **else** 0 **fi**
[] $i > j$ **then if** $i \ge j$ **then** $bin(i, j)$ **else** 0 **fi**
[] $i < j$ **then if** $i \ge j$ **then** $bin(i, j)$ **else** 0 **fi fi**

≡ [ individual simplification of branches ]

<u>case $i = j$</u>

**if** $i \ge j$ **then** $bin(i, j)$ **else** 0 **fi**

≡ [ neutrality of 0 w.r.t. + ]

**if** $i \ge j$ **then** $bin(i, j)$ **else** 0 **fi** + 0

≡ [ $j = i \Rightarrow j-1 = i-1 \land j > i-1$;
simplification of conditional (backwards) ]

**if** $i-1 \geq j-1$ **then** $bin(i-1, j-1)$ **else** 0 **fi** +
**if** $i-1 \geq j$ **then** $bin(i-1, j)$ **else** 0 **fi**

$\equiv$ [ fold *bin'* ]

$bin'(i-1, j-1) + bin'(i-1, j)$

case $i > j$

**if** $i \geq j$ **then** $bin(i, j)$ **else** 0 **fi**

$\equiv$ [ simplification of conditional; unfold *bin* ]

$bin(i-1, j-1) + bin(i-1, j)$

$\equiv$ [ $j < i \Rightarrow j-1 \leq i-1 \wedge j \leq i-1$;
simplification of conditional (backwards) ]

**if** $i-1 \geq j-1$ **then** $bin(i-1, j-1)$ **else** 0 **fi** +
**if** $i-1 \geq j$ **then** $bin(i-1, j)$ **else** 0 **fi**

$\equiv$ [ fold *bin'* ]

$bin'(i-1, j-1) + bin'(i-1, j)$

case $i < j$

**if** $i \geq j$ **then** $bin(i, j)$ **else** 0 **fi**

$\equiv$ [ simplification of conditional; neutrality of 0 w.r.t. + ]

$0 + 0$

$\equiv$ [ $j > i \Rightarrow j-1 > i-1 \wedge j > i-1$;
simplification of conditional (backwards) ]

**if** $i-1 \geq j-1$ **then** $bin(i-1, j-1)$ **else** 0 **fi** +
**if** $i-1 \geq j$ **then** $bin(i-1, j)$ **else** 0 **fi**

$\equiv$ [ fold *bin'* ]

$bin'(i-1, j-1) + bin'(i-1, j)$

**if** $i = j$ **then** $bin'(i-1, j-1) + bin'(i-1, j)$
[] $i > j$ **then** $bin'(i-1, j-1) + bin'(i-1, j)$
[] $i < j$ **then** $bin'(i-1, j-1) + bin'(i-1, j)$ **fi**

$\equiv$ [ simplification of guarded expression ]

$bin'(i-1, j-1) + bin'(i-1, j)$

$bin'(i, j) =_{\text{def}}$ **if** $i = 0$ **then** **if** $j = 0$ **then** 1 **else** 0 **fi**
**else if** $j = 0$ **then** $bin'(i-1, j)$
**else** $bin'(i-1, j-1) + bin'(i-1, j)$ **fi fi**

➤ *Apply property* of (left-)neutrality of 0 w.r.t. + in **else-then**-branch;
*Distributivity of operation over conditional*

*bin'*(*I*, *J*) **where**

*bin'*: (**inat** × **inat**) → **nat**,
*bin'*(*i*, *j*) $=_{def}$ **if** *i* = 0 **then** **if** *j* = 0 **then** 1 **else** 0 **fi**
**else** **if** *j* = 0 **then** 0 **else** *bin'*(*i*–1, *j*–1) **fi** + *bin'*(*i*–1, *j*) **fi**

**2. subgoal**: transformation into an equivalent definition composed of "functional forms"

➤ *Currying* (variant for functions defined by conditional)

*bin''*(*I*)(*J*) **where**

*bin''*: **inat** → (**inat** → **nat**),
*bin''*(*i*) $=_{def}$ **if** *i* = 0 **then** (**inat** *j*) **nat**: **if** *j* = 0 **then** 1 **else** 0 **fi**
**else** (**inat** *j*) **nat**:
**if** *j* = 0 **then** 0 **else** *bin''*(*i*–1)(*j*–1) **fi** + *bin''*(*i*–1)(*j*) **fi**

➤ Conversion into composition of functional forms:

– **then**-branch:

(**inat** *j*) **nat**: **if** *j* = 0 **then** 1 **else** 0 **fi**
≡ [ property of **nat** ]
(**inat** *j*) **nat**: **if** *j* < 1 **then** 1 **else** 0 **fi**
≡ [ fold *SHIFTR* ]
*SHIFTR*((**inat** *j*) **nat**: 0, 1, 1)
≡ [ fold *NEW* ]
*SHIFTR*(*NEW*(0), 1, 1)

– **else**-branch:

(**inat** *j*) **nat**: **if** *j* = 0 **then** 0 **else** *bin''*(*i*–1)(*j*–1) **fi** + *bin''*(*i*–1)(*j*)
≡ [ property of **nat** ]
(**inat** *j*) **nat**: **if** *j* < 1 **then** 0 **else** *bin''*(*i*–1)(*j*–1) **fi** + *bin''*(*i*–1)(*j*)
≡ [ fold *MAP* ]
*MAP*(+, (**inat** *j*) (**nat** × **nat**):
(**if** *j* < 1 **then** 0 **else** *bin''*(*i*–1)(*j*–1) **fi**, *bin''*(*i*–1)(*j*)))
≡ [ fold *ZIP* ]
*MAP*(+, *ZIP*((**inat** *j*) **nat**: **if** *j* < 1 **then** 0 **else** *bin''*(*i*–1)(*j*–1) **fi**, *bin''*(*i*–1))
≡ [ fold *SHIFTR* ]
*MAP*(+, *ZIP*(*SHIFTR*(*bin''*(*i*–1), 1, 0), *bin''*(*i*–1))

*bin''*(*i*) $=_{def}$ **if** *i* = 0 **then** *SHIFTR*(*NEW*(0), 1, 1)
**else** *MAP*(+, *ZIP*(*SHIFTR*(*bin''*(*i*–1), 1, 0), *bin''*(*i*–1)) **fi**

**3. subgoal**: recursion simplification by inverting the computation

➤ *Inversion*

$bin''(i) =_{def}$
$bin'''(i, 0, SHIFTR(NEW(0), 1, 1))$ **where**
$bin'''$: (**inat** × **inat** × (**inat** → **nat**)) → (**inat** → **nat**),
$bin'''(n, i, f) =_{def}$ **if** $i = n$ **then** $f$
**else** $bin'''(n, i{+}1, MAP(+, ZIP(SHIFTR(f, 1, 0), f))$ **fi**

In this sample derivation it took us a lot of elementary steps to reach the first subgoal of a definition of *bin'* with independent parameters. All these, maybe boring, steps have been given on purpose, in order to convince the reader that every tiny step is indeed a formal one. Of course, for practical developments, (part of) the reasoning applied here could be abstracted into corresponding compact transformation rules, if appropriate. Assuming the availability of such a compact rules, we find out that only four transformation steps are sufficient to formally derive the above parallel version of an algorithm computing the binomial coefficient from the usual applicative one.

## 4.3 Algorithm by Cocke, Kasami and Younger

The previous example started from an operational specification of the problem to be solved. Now we want to look at a problem initially specified in a non-operational way. Accordingly, a major part of the development will be devoted to first converting this descriptive specification into an operational one. In particular, we will realize that the remaining steps from the applicative specification to the parallel solution are very much the same as in the previous example. In fact, the underlying strategy is exactly the same.

The problem we want to deal with is a particular recognition algorithm for context-free grammars in Chomsky Normal Form. Before being able to formulate the problem proper, a few preliminaries, such as definitions of the central notions, are needed.

A **context-free grammar** $G$ = (N, T, S, P) is a 4-tuple where:

- N is a finite non-empty set of non-terminal symbols;
- T is a finite non-empty set of terminal symbols with $N \cap T = \varnothing$;
- $S \in N$ is a particular symbol, called "start symbol" or "axiom"; and
- $P \subseteq N \times (N \cup T)^*$ is a finite and non-empty set of "productions".

A context-free grammar $G$ = (N, T, S, P) is in **Chomsky Normal Form**, iff every $p \in P$ is of one of the forms

(a) (A, BC) with A, B, C ∈ N;
(b) (A, a) with A ∈ N, a ∈ T;
(c) (S, <>) provided S does not occur in the righthand side of any other production.

The information given by these definitions can immediately be cast into corresponding type definitions. For our development below, we will use

**t** = ‹ terminal characters ›
**n** = ‹ non-terminal characters ›
**symb** = **t** | **n**
**str** = ESEQU(**symb**)
**tstr** = ESEQU(**t**)
**nstr** = ESEQU(**n**)
**ntstr** = (**tstr** $s$: $s \neq$ <>)
**prod** = PAIR(**n**, **str**)
**nset** = SET(**n**)
**strset** = SET(**str**).

In order to be able to work with somewhat more compact formulations, we also introduce a few auxiliary operations which abbreviate important aspects in connection with grammars.

*"Complex product" of sets of strings:*

.•.: (**strset** × **strset**) → **strset**,
$N \bullet N' =_{def}$ {**str** $v$: ∃ **str** $v'$, **str** $v''$: $v = v'v'' \wedge v' \in N \wedge v'' \in N'$},

*Lefthand sides of a set of strings* (with respect to a given grammar):

*lhs*: **strset** → **nset**,
$lhs(N) =_{def}$ {**n** $u$: ∃ **str** $n$: $n \in N \wedge (u, n) \in$ P},

*Lefthand sides of a string* (with respect to a given grammar):

*lhs'*: **str** → **nset**,
$lhs'(s) =_{def}$ {**n** $u$: $(u, s) \in$ P},

*Strings as mappings from indices to symbols:*

*str*: **str** → (**nat** → **symb**),
$str(s) =_{def}$ (**nat** $i$: $1 \leq i \leq |s|$) **symb**: $s[i]$

Grammars are used to define languages as sets of terminal strings derivable from the start symbol of the grammar. Derivability of a string $y$ from a string $x$ is defined by

$x \to^* y \ =_{def} (x = y) \triangledown (\exists$ **str** $z$: $x \to z \wedge z \to^* y)$

where

$x \rightarrow y =_{def} \exists$ **str** $l, r,$ **prod** $(a, b) \in P$: $x = lar \wedge y = lbr$.

With respect to derivability, many additional properties can be proved. In our subsequent development we will only need two of these properties (for **n** $x, x'$, **ntstr** $w$), viz.

$$x \rightarrow^{*} w \equiv x \in \{\mathbf{n}\ u: u \rightarrow^{*} w\} \qquad (4.3\text{-}1)$$

and

$$xx' \rightarrow^{*} w \equiv \exists\ \mathbf{ntstr}\ w', w'': w = w'w'' \wedge x \rightarrow^{*} w' \wedge x' \rightarrow^{*} w'' \qquad (4.3\text{-}2)$$

where the latter property may be considered as an alternative definition of context-freeness.

Now we are in a position to formally specify the recognition problem (for **ntstr** $W$ and grammar $G$ in Chomsky Normal Form):

$RP(W)$ **where**

$RP$: **ntstr** $\rightarrow$ **bool**,
$RP(w) =_{def} S \rightarrow^{*} w$

For shortening the presentation of the subsequent derivation we assume that the given grammar has only productions of the forms (a) and (b). This is not a true restriction, since productions of form (c) simply can be handled by an additional initial case distinction. Furthermore, we use $W$ and $G$ as global parameters.

**general goal**: transformation into a solution composed of functional forms
**1. subgoal**: derivation of an applicative solution

➤ *Apply property* (4.3-1); *Abstraction*:

$RP(w) =_{def} S \in CKY(w)$ **where**

$CKY$: **ntstr** $\rightarrow$ **nset**,
$CKY(w) =_{def} \{\mathbf{n}\ u: u \rightarrow^{*} w\}$

➤ *Unfold* definition $\rightarrow^{*}$:

$CKY(w) =_{def} \{\mathbf{n}\ u: (u = w) \vee (\exists\ \mathbf{str}\ v: u \rightarrow v \wedge v \rightarrow^{*} w)\}$

➤ *Simplification* ($(u = w) \equiv$ **false**, since $u \in$ **n**, $w \in$ **ntstr**)

$CKY(w) =_{def} \{\mathbf{n}\ u: \exists\ \mathbf{str}\ v: u \rightarrow v \wedge v \rightarrow^{*} w\}$

➤ *Case introduction* ($|w| = 1 \vee |w| > 1$)

$CKY(w) =_{def}$ **if** $|w| = 1$ **then** $\{\mathbf{n}\ u: \exists\ \mathbf{str}\ v: u \rightarrow v \wedge v \rightarrow^{*} w\}$
**else** $\{\mathbf{n}\ u: \exists\ \mathbf{str}\ v: u \rightarrow v \wedge v \rightarrow^{*} w\}$ **fi**

➤ *Simplifications* and *Rearrangements*

– simplification in **then**-branch (under premise $|w| = 1$)

$\{\mathbf{n}\ u: \exists\ \mathbf{str}\ v: u \to v \wedge v \to^* w\}$

$\equiv$ [ Chomsky Normal Form ]

$\{\mathbf{n}\ u: u \to w\}$

$\equiv$ [ definition of $\to$ ; premise]

$\{\mathbf{n}\ u: (u, w) \in \mathrm{P}\}$

$\equiv$ [ fold *lhs'* ]

$lhs'(w)$

– simplification in **else**-branch (under premise $|w| > 1$)

$\{\mathbf{n}\ u: \exists\ \mathbf{str}\ v: u \to v \wedge v \to^* w\}$

$\equiv$ [ $u \in \mathbf{n}$, Chomsky Normal Form ]

$\{\mathbf{n}\ u: \exists\ \mathbf{n}\ v, v': (u, vv') \in \mathrm{P} \wedge vv' \to^* w\}$

$\equiv$ [ property (4.3-2) of context-free grammars ]

$\{\mathbf{n}\ u: \exists\ \mathbf{n}\ v, v': (u, vv') \in \mathrm{P} \wedge$
$\exists\ \mathbf{ntstr}\ w', w'': w = w'w'' \wedge v \to^* w' \wedge v' \to^* w''\}$

$\equiv$ [ set properties ]

$\bigcup_{\mathbf{ntstr}\ w', w'':\ w = w'w''} \{\mathbf{n}\ u: \exists\ \mathbf{n}\ v, v': (u, vv') \in \mathrm{P} \wedge$
$v \to^* w' \wedge v' \to^* w''\}$

$\equiv$ [ property (4.3-1) ]

$\bigcup_{\mathbf{ntstr}\ w', w'':\ w = w'w''} \{\mathbf{n}\ u: \exists\ \mathbf{n}\ v, v': (u, vv') \in \mathrm{P} \wedge$
$v \in \{\mathbf{n}\ z: z \to^* w'\} \wedge v' \in \{\mathbf{n}\ z: z \to^* w''\}\}$

$\equiv$ [ fold • ]

$\bigcup_{\mathbf{ntstr}\ w', w'':\ w = w'w''} \{\mathbf{n}\ u: \exists\ \mathbf{str}\ v'': (u, v'') \in \mathrm{P} \wedge$
$v'' \in (\{\mathbf{n}\ z: z \to^* w'\} \bullet \{\mathbf{n}\ z: z \to^* w''\})\}$

$\equiv$ [ fold *lhs* ]

$\bigcup_{\mathbf{ntstr}\ w', w'':\ w = w'w''} lhs(\{\mathbf{n}\ z: z \to^* w'\} \bullet \{\mathbf{n}\ z: z \to^* w''\})$

$CKY(w) =_{\mathrm{def}}$ **if** $|w| = 1$ **then** $lhs'(w)$
**else** $\bigcup_{\mathbf{ntstr}\ w', w'':\ w = w'w''} lhs(\{\mathbf{n}\ z: z \to^* w'\} \bullet$
$\{\mathbf{n}\ z: z \to^* w''\})$ **fi**

➤ Fold *CKY*

$\mathrm{S} \in CKY(w)$ **where**

$CKY$: **ntstr** $\to$ **nset**,

$CKY(w) =_{\mathrm{def}}$ **if** $|w| = 1$ **then** $lhs'(w)$
**else** $\bigcup_{\mathbf{ntstr}\ w', w'':\ w = w'w''} lhs(CKY(w') \bullet CKY(w''))$ **fi**

**2. subgoal**: improvement of efficiency

➤ *Data type representation* (indices $i, j$ of type **inat** = (**nat** $i$: $1 \le i \le |W|$) instead of **ntstr** $w$, cf. [8]) using the assertion:

$\forall$ **ntstr** $w$: $\exists$ **inat** $i, j$: $1 \le i \le |W|-j+1 \;\Delta\; w = W[i{:}i+j-1]$

- $W \;\equiv\; (1, |W|)$
- $|w| = 1 \;\equiv\; j = 1$
- $lhs'(w) \;\equiv\; lhs'(W[i])$, if $|w| = 1$
- for set expressions Q:

  $\bigcup_{\mathbf{ntstr}\; w', w'':\, w = w'w''} Q(w', w'') \equiv$
  $\bigcup_{1 \le k \le j-1} Q(W[i{:}i+k-1], W[i+k{:}i+j-1])$

S $\in$ $CKY(1, |W|)$ **where**

$CKY$: (**inat** $\times$ **inat**) $\rightarrow$ **nset**,
**def**($CKY(i, j)$) $\Rightarrow$ $1 \le i \le |W|-j+1$,
$CKY(i, j) =_{def}$ **if** $j = 1$ **then** $lhs'(W[i])$
**else** $\bigcup_{1 \le k \le j-1} lhs(CKY(i, k) \bullet CKY(i+k, j-k))$ **fi**

**3. subgoal**: transformation into an equivalent definition with independent parameters

➤ New version with independent parameters
(through totalization, as in section 4.2)

S $\in$ $CKY(1, |W|)$ **where**

$CKY$: (**inat** $\times$ **inat**) $\rightarrow$ **nset**,
$CKY(i, j) =_{def}$ **if** $j = 1$ **then** $lhs'(W[i])$
**else** $\bigcup_{1 \le k \le j-1} lhs(CKY(i, k) \bullet$
**if** $i > |W|-k$ **then** $\varnothing$ **else** $CKY(i+k, j-k)$ **fi**)) **fi**

**4. subgoal**: transformation into equivalent definition composed of functional forms

➤ *Currying* (on second argument, variant for functions defined by conditionals)

S $\in$ $CKY'(|W|)(1)$ **where**

$CKY'$: **inat** $\to$ (**inat** $\to$ **nset**),
$CKY'(j) =_{\text{def}}$ **if** $j = 1$ **then** (**inat** $i$) **nset**: $lhs'(W[i])$
**else** (**inat** $i$) **nset**:
$\bigcup_{1 \le k \le j-1} lhs(CKY'(k)(i) \bullet$
**if** $i > |W|-k$ **then** $\varnothing$ **else** $CKY'(j-k)(i+k)$ **fi**) **fi**

➤ *Introduction* of an auxiliary functional (through "lifting");
*Distributivity* of abstraction over union

$CKY'(j) =_{\text{def}}$ **if** $j = 1$ **then** (**inat** $i$) **nset**: $lhs'(W[i])$
**else** $\bigcup_{1 \le k \le j-1}$
(**inat** $i$) **nset**: $lhs(\ CKY'(k)(i) \bullet$
**if** $i > |W|-k$ **then** $\varnothing$ **else** $CKY'(j-k)(i+k)$ **fi**) **fi**

**where**

$\bigcup_{1 \le k \le j-1}$: (**inat** $\to$ **nset**) $\to$ (**inat** $\to$ **nset**),
$\bigcup_{1 \le k \le j-1} f =_{\text{def}}$ (**inat** $i$) **nset**: $\bigcup_{1 \le k \le j-1} f(i)$

➤ *Conversion* into composition of functional forms:

- **then**-branch:

(**inat** $i$) **nset**: $lhs'(W[i])$
≡ [ fold *MAP* ]
$MAP(lhs'$, (**inat** $i$) **str**: $W[i])$
≡ [ fold *str* ]
$MAP(lhs', str(W))$

- **else**-branch:

(**inat** $i$) **nset**: $lhs(CKY'(k)(i) \bullet$
**if** $i > |W|-k$ **then** $\varnothing$ **else** $CKY'(j-k)(i+k)$ **fi**)
≡ [ fold *MAP* ]
$MAP(lhs$, (**inat** $i$) (**nset**): $CKY'(k)(i) \bullet$
**if** $i > |W|-k$ **then** $\varnothing$ **else** $CKY'(j-k)(i+k)$ **fi**)
≡ [ fold *MAP* ]
$MAP(lhs, MAP(\bullet$, (**inat** $i$) (**nset** × **nset**):
$(CKY'(k)(i)$, **if** $i > |W|-k$ **then** $\varnothing$ **else** $CKY'(j-k)(i+k)$ **fi**)))
≡ [ fold *ZIP* ]
$MAP(lhs, MAP(\bullet, ZIP(CKY'(k)$, (**inat** $i$) **nset**:
**if** $i > |W|-k$ **then** $\varnothing$ **else** $CKY'(j-k)(i+k)$ **fi**)))
≡ [ fold *SHIFTL* ]
$MAP(lhs, MAP(\bullet, ZIP(CKY'(k), SHIFTL(CKY'(j-k), k, \varnothing))))$

$S \in CKY'(|W|)(1)$ **where**

$CKY'$: **inat** $\rightarrow$ (**inat** $\rightarrow$ **nset**),
$CKY'(j) =_{def}$ **if** $j = 1$ **then** $MAP(lhs', str(W))$
**else** $\bigcup_{1 \le k \le j-1} MAP(lhs', MAP(\bullet, ZIP(CKY'(k),$
$SHIFTL(CKY'(j-k), k, \varnothing))))$ **fi**

**5. subgoal**: simplification of recursion

➤ *Tabulation* (cf. [8])

$S \in CKY''(2, MAP(lhs', str(W)))(|W|)$ **where**

$CKY''$: (**inat** $\times$ (**inat** $\rightarrow$ (**inat** $\rightarrow$ **nset**))) $\rightarrow$ (**inat** $\rightarrow$ **nset**),
**def**$(CKY''(j, T)) \Rightarrow \forall$ (**inat** $j'$: $j' < j$): $T(j') = CKY'(j')$,
$CKY''(j, T) =_{def}$
**if** $j > |W|$ **then** $T$
**else** $CKY''(j+1, EXT(T, 1, \bigcup_{1 \le k \le j-1} MAP(lhs, MAP(\bullet, ZIP(T(k),$
$SHIFTL(T(j-k), k, \varnothing)))))$ **fi**

A fully operational solution to our problem needs at least an operational definition of $\bigcup_{1 \le k \le j-1}$. Since, however, this best can be done coupled with a transition from the recursive formulation to an iterative one, we omit this straightforward final step.

## 4.4 Sorting

As a last example to demonstrate transforming towards parallel executability, we consider the problem of sorting an array $S$ (with $|S| \ge 2$) of type (**inat** $\rightarrow$ **m**) which can be specified by:

$sort(S)$ **where**

**inat** = (**nat** $i$: $1 \le i \le |S|$),

$sort$: (**inat** $\rightarrow$ **m**) $\rightarrow$ (**inat** $\rightarrow$ **m**),
$sort(s) =_{def}$ **some** (**inat** $\rightarrow$ **m**) $x$: $hassameels(x, s) \wedge issorted(x)$,

$hassameels$: ((**inat** $\rightarrow$ **m**) $\times$ (**inat** $\rightarrow$ **m**)) $\rightarrow$ **bool**,
$hassameels(x, s) =_{def}$ ‹‹ the bags of images of $x$ and $s$ are the same ››,

$issorted$: (**inat** $\rightarrow$ **m**) $\rightarrow$ **bool**,
$issorted(s) =_{def} \#invs(s) = 0$,

$\#invs$: (**inat** $\rightarrow$ **m**) $\rightarrow$ **nat**,
$\#invs(s) =_{def} |\{($**inat** $i$, **inat** $j)$: $i < j \wedge s(i) > s(j)\}|$.

The purpose of this example is to demonstrate that under certain circumstances a direct transition from descriptive, non-operational constructs to skeletons is possible and also feasible.

Algorithms for *sort* can be derived in various ways (cf. e.g. [2]). One key idea used in such derivations is to synthesize an algorithm where in each step of the computation the number of inversions is reduced, e.g., by swapping elements. Since, for $n = LEN(s)$, $n^2/2$ is an upper bound for the number of inversions, a naive algorithm (which removes one inversion per computation step) would need $n^2/2$ steps in the worst case. A substantial improvement would be achieved, if the maximum number of computation steps could be reduced to $n$. This, however, would require an algorithm where the number of inversions after computation step $i$ is at most $n(n-i)/2$. In order to formally derive such an algorithm from our original specification, we generalize the original function *sort* to a new function *sort'* with two additional arguments (a "step counter" $i$ and an upper bound $n$ for the number of steps) and an assertion that characterizes the intended behaviour. Technically we proceed as follows (where $|s|$ is as an abbreviation for $LEN(s)$):

➤ *Embedding* with assertion

$sort(s) =_{def} sort'(s, 0, |s|)$ **where**

$sort'$: ((**inat** → **m**) × **nat** × **inat**) → (**inat** → **m**),
**defined**($sort'(s, i, n)$) $\Rightarrow n = |s| \Delta \#invs(s) \le n(n-i)/2$,
$sort'(s, i, n) =_{def} sort(s)$

➤ *Unfold sort*; *Case introduction* ($i \ge n \vee i < n$) and *Distributivity* of comprehensive choice over conditional

$sort'(s, i, n) =_{def}$
**if** $i \ge n$ **then some** (**inat** → **m**) $x$: $hassameels(x, s) \wedge issorted(x)$
**else some** (**inat** → **m**) $x$: $hassameels(x, s) \wedge issorted(x)$ **fi**

➤ *Simplifications*

– **then**-branch (under premise $\#invs(s) \le n(n-i)/2 \wedge i \ge n$):
$(\#invs(s) \le n(n-i)/2 \wedge i \ge n) \Rightarrow \#invs(s) \le 0 \Leftrightarrow issorted(s)$;
$hassameels(s, s) \equiv$ **true**

– **else**-branch (under premise $\#invs(s) \le n(n-i)/2 \wedge i < n$):
find operation
*transp*: (**inat** → **m**) → (**inat** → **m**),
with properties
$s = (transp(S))^i \Rightarrow (\#invs(transp(s)) \le n(n-i-2)/2)$; (4.4-1)
$hassameels(transp(s), s) \equiv$ **true**

$sort'(s, i, n) =_{def}$
  **if** $i \geq n$ **then** $s$
    **else some (inat → m)** $x$: $hassameels(x, transp(s)) \wedge issorted(x)$ **fi**

➤ *Folding* (with assertion)

$sort'(s, i, n) =_{def}$ **if** $i \geq n$ **then** $s$ **else** $sort'(transp(s), i+2, n)$ **fi**

In order for this definition to be truly operational, we still have to supply an explicit definition of the operation *transp* the properties of which have been stated in (4.4-1). Of course, these properties provide a valuable guide-line in finding such a definition. Nevertheless, this step requires intuition and, thus, is a major Eureka step. The following definition can be shown (cf. [12]) to satisfy the required properties:

*transp*: **(inat → m) → (inat → m)**,
$transp(s) =_{def} transpe(transpo(s))$ **where**

  **inath** = (**inat** $i$: $i \leq (|S|$ **div** $2)$),
  *transpo*: **(inat → m) → (inat → m)**,
  $transpo(s) =_{def}$ **that (inat → m)** $b$: (4.4-2)
   (**even** $|s| \wedge \forall$ **inath** $i$: $(b(2i-1), b(2i)) = mm(s(2i-1), s(2i))) \nabla$
   (**odd** $|s| \wedge \forall$ **inath** $i$: $(b(2i-1), b(2i)) = mm(s(2i-1), s(2i)) \wedge b(|s|) = s(|s|))$,

  **inathh** = (**inat** $i$: $i \leq ((|S|-1)$ **div** $2)$),
  *transpe*: **(inat → m) → (inat → m)**,
  $transpe(s) =_{def}$ **that (inat → m)** $b$: (4.4-3)
   (**even** $|s| \wedge \forall$ **inathh** $i$: $(b(2i), b(2i+1)) = mm(s(2i), s(2i+1)) \wedge$
     $b(1) = s(1) \wedge b(|s|) = s(|s|)) \nabla$
   (**odd** $|s| \wedge \forall$ **inathh** $i$: $(b(2i), b(2i+1)) = mm(s(2i), s(2i+1)) \wedge$
     $b(1) = s(1))$,

  *mm*: **(m × m) → (m × m)**,
  $mm(x, y) =_{def}$ **if** $x \leq y$ **then** $(x, y)$ [] $x \geq y$ **then** $(y, x)$ **fi**.

This algorithm is a high-level description of a sorting algorithm known as "odd-even transposition sort" (cf. [1], [6], [10]).

Since (equivalent) operational equivalents of these definitions are straightforward, we are basically done, when aiming at a sequential algorithm. In order to transform to parallel executability, however, as well as for improving the sequential algorithm, we perform a few more transformational steps all of which are essentially to be considered as data type transformations.

➤ *Data type transformation*: adding "fictitious elements"

In order to avoid within *transpo* and *transpe* the case distinctions w.r.t. the length of $s$ being odd or even, we use a *data type embedding* which extends the domain of the array involved using a "fictitious" element **d** on **m** and "lifts" all functions involved to the extended type. To this end we define

$$\mathbf{m}^+ = (\mathbf{m} \mid \mathbf{d}),$$
$$\mathbf{inat}^+ = (\mathbf{nat}\ i\colon 1 \le i \le 2((|S|\ \mathbf{div}\ 2)+1)),$$

redefine **inat** (equivalently) by

$$\mathbf{inat} = (\mathbf{inat}^+\ i\colon i \le |S|),$$

and use arrays of type $(\mathbf{inat}^+ \rightarrow \mathbf{m}^+)$ instead of those of type $(\mathbf{inat} \rightarrow \mathbf{m})$. Intuitively, this implementation means that we add some fictitious elements (one in case of odd length, two in case of even length) at the high end of the arrays.

For defining the "lifting" of functions we introduce auxiliary functionals (that characterize the "implementation" and the "abstraction" mappings, cf. [8])

$$.^+\colon (\mathbf{inat} \rightarrow \mathbf{m}) \rightarrow (\mathbf{inat}^+ \rightarrow \mathbf{m}^+),$$
$$s^+ =_{\text{def}} (\mathbf{inat}^+\ i)\ \mathbf{m}^+\colon \mathbf{if}\ i\ \mathbf{is}\ \mathbf{inat}\ \mathbf{then}\ s(i)\ \mathbf{else}\ \mathbf{d}\ \mathbf{fi},$$

$$.^-\colon (\mathbf{inat}^+ \rightarrow \mathbf{m}^+) \rightarrow (\mathbf{inat} \rightarrow \mathbf{m}),$$
$$\mathbf{defined}(t^-) \Rightarrow \forall\ \mathbf{inat}\ i\colon \mathbf{if}\ i\ \mathbf{is}\ \mathbf{inat}\ \mathbf{then}\ t(i)\ \mathbf{is}\ \mathbf{m}\ \mathbf{fi},$$
$$t^- =_{\text{def}} (\mathbf{inat}\ \ i)\ \mathbf{m}\colon t(i).$$

Obviously, these functionals are injective and also inverses of each other: For $s$ and $t$ of appropriate type, the proofs of the properties

$$(t^-)^+ \equiv t, \text{ and}$$
$$(s^+)^- \equiv s$$

are straightforward.

Moreover, also obviously,

$$\mathbf{even}\ |s^+| \quad \text{and} \quad \mathbf{last}(s^+)\ \mathbf{is}\ \mathbf{d} \tag{4.4-4}$$

hold for arbitrary $s$ of type $(\mathbf{inat} \rightarrow \mathbf{m})$. This will be used below to get rid of the case distinctions in (4.4-2) and (4.4-3).

For formally deriving the lifted versions of the functions involved, we first introduce

$$sort''\colon ((\mathbf{inat}^+ \rightarrow \mathbf{m}^+) \times \mathbf{nat} \times \mathbf{inat}) \rightarrow (\mathbf{inat}^+ \rightarrow \mathbf{m}^+),$$
$$\mathbf{defined}(sort''(s, i, n)) \Rightarrow isext(s) \land n = |s^-| \land \#invs(s^-) \le n(n-i)/2,$$
$$sort''(s, i, n) =_{\text{def}} (sort'(s^-, i, n))^+,$$

(where the assertion of $sort''$ follows immediately from its definition as a function composition and the assertions of its constituents).

A definition of *sort''* which is independent of the definition of *sort'* can be calculated (according to the "function composition strategy" in [8]) as follows:

$sort''(s, i, n)$
$\equiv$ [ unfold $sort''$ ]
$(sort'(s^-, i, n))^+$
$\equiv$ [ unfold $sort'$ ]
$(\textbf{if } i \geq n \textbf{ then } s^- \textbf{ else } sort'(transp(s^-), i+2, n) \textbf{ fi})^+$
$\equiv$ [ distributivity of operations over conditional ]
$\textbf{if } i \geq n \textbf{ then } (s^-)^+ \textbf{ else } (sort'(transp(s^-), i+2, n))^+ \textbf{ fi}$
$\equiv$ [ above properties of functionals ]
$\textbf{if } i \geq n \textbf{ then } s \textbf{ else } (sort'(((transp(s^-))^+)^-, i+2, n))^+ \textbf{ fi}$
$\equiv$ [ introduction of a new function $transp^+$ defined by
$transp^+(s) =_{def} (transp(s^-))^+$ ] (4.4-5)
$\textbf{if } i \geq n \textbf{ then } s \textbf{ else } (sort'((transp^+(s))^-, i+2, n))^+ \textbf{ fi}$
$\equiv$ [ folding ]
$\textbf{if } i \geq n \textbf{ then } s \textbf{ else } sort''(transp^+(s), i+2, n) \textbf{ fi}.$

Analogously, we calculate from (4.4-5), i.e. from

$transp^+(s) =_{def} (transp(s^-))^+,$

a definition of $transp^+$ which is independent of *transp*. The same procedure is then applied to introduce new functions $transpo^+$, $transpe^+$ and $mm^+$ and to derive new independent definitions for them.

Thus, altogether, we get

$sort(s) =_{def} (sort''(s+, 0, n))^-$ **where**

$sort''$: $((\mathbf{inat^+} \to \mathbf{m^+}) \times \mathbf{nat} \times \mathbf{inat}) \to (\mathbf{inat^+} \to \mathbf{m^+})$,
$\textbf{defined}(sort''(s, i, n)) \Rightarrow isext(s) \,\Delta\, n = |s^-| \,\Delta\, \#invs(s^-) \leq n(n-i)/2$,
$sort''(s, i, n) =_{def}$
$\textbf{if } i \geq n \textbf{ then } s \textbf{ else } sort''(transp^+(s), i+2, n) \textbf{ fi}$ **where**

$transp^+$: $(\mathbf{inat^+} \to \mathbf{m^+}) \to (\mathbf{inat^+} \to \mathbf{m^+})$,
$transp^+(s) =_{def} transpe^+(transpo^+(s))$ **where**

$\mathbf{inath^+} = (\mathbf{inat^+}\ i\text{: } i \leq (|S| \textbf{ div } 2)+1)$,
$transpo^+$: $(\mathbf{inat^+} \to \mathbf{m^+}) \to (\mathbf{inat^+} \to \mathbf{m^+})$,
$transpo^+(s) =_{def}$ **that** $(\mathbf{inat^+} \to \mathbf{m^+})\ b$:
$\forall\, \mathbf{inath^+}\ i\text{: } (b(2i-1), b(2i)) = mm^+(s(2i-1), s(2i))$

$\mathbf{inathh^+} = (\mathbf{inat^+}\ i\text{: } i \leq (|S| \textbf{ div } 2))$,
$transpe^+$: $(\mathbf{inat^+} \to \mathbf{m^+}) \to (\mathbf{inat^+} \to \mathbf{m^+})$,

$$transpe^{+}(s) =_{def} \textbf{that } (\textbf{inat}^{+} \to \textbf{m}^{+})\ b: \qquad (4.4\text{-}6)$$
$$\forall\ \textbf{inathh}^{+}\ i: \quad (b(2i), b(2i{+}1)) = mm^{+}(s(2i), s(2i{+}1)) \wedge$$
$$b(1) = s(1) \wedge b(|s|) = s(|s|),$$

$$mm^{+}: (\textbf{m}^{+} \times \textbf{m}^{+}) \to (\textbf{m}^{+} \times \textbf{m}^{+}),$$
$$mm^{+}(x, y) =_{def} \textbf{if } (x, y) \textbf{ is } (\textbf{m} \times \textbf{m}) \textbf{ then } mm(x, y) \textbf{ else } (x, y) \textbf{ fi}$$

➤ *Local change* of the definition of $transpe^{+}$

Next, in order to obtain a closed form of the quantified formula in (4.4-6), we extend the domain of the universal quantifier (by using a simple index translation) and introduce operations

$$d^{+}: (\textbf{inat}^{+} \to \textbf{m}^{+}) \to (\textbf{inat}^{+} \to \textbf{m}^{+}),$$
$$d^{+}(s) =_{def} (\textbf{inat}^{+}\ i)\ \textbf{m}^{+}: \textbf{if } i < 2 \textbf{ then d else } s(i-1) \textbf{ fi},$$

$$d^{-}: (\textbf{inat}^{+} \to \textbf{m}^{+}) \to (\textbf{inat}^{+} \to \textbf{m}^{+}),$$
$$d^{-}(s) =_{def} (\textbf{inat}^{+}\ i)\ \textbf{m}^{+}: \textbf{if } i > (|S| - 1) \textbf{ then d else } s(i+1) \textbf{ fi}.$$

Using (4.4-4) we can prove for s of type (**inat** → **m**)

$$d^{-}(d^{+}(s^{+})) \equiv s^{+}. \qquad (4.4\text{-}7)$$

From this we can straightforwardly derive

$$transpe^{+}(s) \equiv\ d^{-}(\textbf{that } (\textbf{inat}^{+} \to \textbf{m}^{+})\ b:$$
$$\forall\ \textbf{inath}^{+}\ i: \quad (b(2i-1), b(2i)) = mm^{+}(d^{+}(s)(2i-1), d^{+}(s)(2i))),$$

such that $transpe^{+}$ can now be expressed in terms of $transpo^{+}$:

$$transpe^{+}(s) \equiv d^{-}(transpo^{+}(d^{+}(s))).$$

Thus, accordingly, $transp^{+}$ can be redefined into

$$transp^{+}(s) =_{def} d^{-}(transpo^{+}(d^{+}(transpo^{+}(s)))).$$

➤ *Data type transformation*: representing an array by a pair of arrays

As a next step we try to treat odd and even indices of the argument $s$ of $sort''$ separately. To this end we use again a data type transformation that splits $s$ of type $\textbf{inat}^{+} \to \textbf{m}^{+}$ into two parts, $o$ and $e$, both of type $\textbf{inath}^{+} \to \textbf{m}^{+}$. Formally (cf. [8] for a detailed treatment of this technique) this data type transformation is based on the assertion

$$s = merge(o, e)\ \textbf{where}$$
$$merge: ((\textbf{inath}^{+} \to \textbf{m}^{+}) \times (\textbf{inath}^{+} \to \textbf{m}^{+})) \to (\textbf{inat}^{+} \to \textbf{m}^{+}),$$
$$merge(o, e) =_{def}$$
$$\textbf{that } (\textbf{inat}^{+} \to \textbf{m}^{+})\ b: \forall\ \textbf{inath}^{+}\ i: \quad (b(2i-1), b(2i)) = (o(i), e(i))$$

which formally maintains the relationship between the original data structure and its representation.

From the definitions of $d^+$, $d^-$ and *merge* it is straightforwardly to prove that

$$d^+(merge(o, e)) \equiv merge(d^+(e), o) \tag{4.4-8}$$

and

$$d^-(merge(o, e)) \equiv merge(e, d^-(o)) \tag{4.4-9}$$

hold for all $o$ and $e$ of appropriate type.

For splitting $s$, we use

$split$: $(\mathbf{inat}^+ \to \mathbf{m}^+) \to ((\mathbf{inath}^+ \to \mathbf{m}^+) \times (\mathbf{inath}^+ \to \mathbf{m}^+))$,

defined by

$split(s) =_{\mathrm{def}}$ **that** $((\mathbf{inath}^+ \to \mathbf{m}^+) \times (\mathbf{inath}^+ \to \mathbf{m}^+))\ (o, e)$: $merge(o, e) = s$.

Obviously, both *merge* and *split* are injective and satisfy (for $s$ of type $\mathbf{inat} \to \mathbf{m}$ and $o, e$ of type $\mathbf{inath}^+ \to \mathbf{m}^+$)

$$merge(split(s^+)) \equiv s^+ \tag{4.4-10}$$
$$split(merge(o, e)) \equiv (o, e). \tag{4.4-11}$$

Next we introduce

$sort^*$: $((\mathbf{inath}^+ \to \mathbf{m}^+) \times (\mathbf{inath}^+ \to \mathbf{m}^+) \times \mathbf{nat} \times \mathbf{inat}) \to (\mathbf{inat}^+ \to \mathbf{m}^+)$

with the assertion

$$\mathbf{defined}(sort^*(o, e, i, n)) \Rightarrow isext(merge(o, e)) \wedge n = |(merge(o, e))^-| \wedge \#invs((merge(o, e))^-) \leq n(n-i)/2,$$

defined by

$sort^*(o, e, i, n) =_{\mathrm{def}} sort''(merge(o, e), i, n)$.

Using (4.4-10) we obtain for the original call of $s$

$sort(s) =_{\mathrm{def}} (sort^*(split(s^+), 0, n))^-$.

A definition of $sort^*$ which is independent of $sort''$ can be calculated as follows:

$sort^*(o, e, i, n)$
  $\equiv$ [ unfold $sort^*$ ]
$sort''(merge(o, e), i, n)$
  $\equiv$ [ unfold $sort''$ ]
**if** $i \geq n$ **then** $merge(o, e)$ **else** $sort''(transp^+(merge(o, e)), i+2, n)$ **fi**

$\equiv$ [ use *transp** defined by
*merge*(*transp**(*o*, *e*)) $=_{\mathrm{def}}$ *transp*$^+$(*merge*(*o*, *e*)) ] (4.4-12)

**if** $i \geq n$ **then** *merge*(*o*, *e*) **else** *sort*''(*merge*(*transp**(*o*, *e*)), *i*+2, *n*) **fi**

$\equiv$ [ fold *sort** ]

**if** $i \geq n$ **then** *merge*(*o*, *e*) **else** *sort**(*transp**(*o*, *e*)), *i*+2, *n*) **fi**

Note that (4.4-12) is indeed a sound definition of *transp** due to injectivity of *split* and (4.4-11).

Likewise, we can calculate a definition of *transp**:

*transp**(*o*, *e*)
$\equiv$ [ (4.4-11) ]
*split*(*merge*(*transp**(*o*, *e*)))
$\equiv$ [ (4.4-12) ]
*split*(*transp*$^+$(*merge*(*o*, *e*)))
$\equiv$ [ unfold *transp*$^+$ ]
*split*($d^-$(*transpo*$^+$($d^+$(*transpo*$^+$(*merge*(*o*, *e*))))))
$\equiv$ [ introduce *transpo** defined by
*merge*(*transpo**(*o*, *e*)) $=_{\mathrm{def}}$ *transpo*$^+$(*merge*(*o*, *e*)) ] (4.4-13)
*split*($d^-$(*transpo*$^+$($d^+$(*merge*(*transpo**(*o*, *e*))))))
$\equiv$ [ introduction of auxiliary variables ]
*split*($d^-$(*transpo*$^+$($d^+$(*merge*(*o'*, *e'*))))) **where** (*o'*, *e'*) = *transpo**(*o*, *e*)
$\equiv$ [ (4.4-8) ]
*split*($d^-$(*transpo*$^+$(*merge*($d^+$(*e'*), *o'*)))) **where** (*o'*, *e'*) = *transpo**(*o*, *e*)
$\equiv$ [ (4.4-13) ]
*split*($d^-$(*merge*(*transpo**($d^+$(*e'*), *o'*)))) **where** (*o'*, *e'*) = *transpo**(*o*, *e*)
$\equiv$ [ introduction of auxiliary variables ]
*split*($d^-$(*merge*(*o''*, *e''*))) **where**
(*o'*, *e'*) = *transpo**(*o*, *e*); (*o''*, *e''*) = *transpo**($d^+$(*e'*), *o'*)
$\equiv$ [ (4.4-9) ]
*split*(*merge*(*e''*, $d^-$(*o''*))) **where**
(*o'*, *e'*) = *transpo**(*o*, *e*); (*o''*, *e''*) = *transpo**($d^+$(*e'*), *o'*)
$\equiv$ [ (4.4-11) ]
(*e''*, $d^-$(*o''*)) **where** (*o'*, *e'*) = *transpo**(*o*, *e*); (*o''*, *e''*) = *transpo**($d^+$(*e'*), *o'*).

Finally, in the same way, a definition of *transpo** is calculated:

*transpo**(*o*, *e*)
$\equiv$ [ (4.4-11) ]
*split*(*merge*(*transpo**(*o*, *e*)))
$\equiv$ [ (4.4-13) ]

$split(transpo^+(merge(o, e)))$

$\equiv$ [ unfold $transpo^+$ ]

$split(\textbf{that } (\textbf{inat}^+ \to \textbf{m}^+)\ b:$
$\quad \forall\ \textbf{inath}^+\ i:\ (b(2i-1), b(2i)) = mm^+(merge(o, e)(2i-1), merge(o, e)(2i)))$

$\equiv$ [ $merge(o, e)(2i-1) \equiv o(i)$; $merge(o, e)(2i) \equiv e(i)$ ]

$split(\textbf{that } (\textbf{inat}^+ \to \textbf{m}^+)\ b: \forall\ \textbf{inath}^+\ i:\ (b(2i-1), b(2i)) = mm^+(o(i), e(i)))$

$\equiv$ [ unfold *split* and simplification ]

$\textbf{that } ((\textbf{inath}^+ \to \textbf{m}^+) o', (\textbf{inath}^+ \to \textbf{m}^+)\ e'):$
$\quad \forall\ \textbf{inath}^+\ i:\ (o'(i), e'(i)) = mm^+(o(i), e(i)))$

$\equiv$ [ splitting a pair ]

$(\textbf{that } (\textbf{inath}^+ \to \textbf{m}^+)\ o': \forall\ \textbf{inath}^+\ i:\ o'(i) = (mm^+(o(i), e(i))).1,$
$\ \textbf{that } (\textbf{inath}^+ \to \textbf{m}^+)\ e': \forall\ \textbf{inath}^+\ i:\ e'(i) = (mm^+(o(i), e(i))).2)$

$\equiv$ [ abstraction ]

$((\textbf{inath}^+\ i:)\ \textbf{m}^+: (mm^+(o(i), e(i))).1,\ (\textbf{inath}^+\ i:)\ \textbf{m}^+: (mm^+(o(i), e(i))).2).$

As a result, we obtain

$sort(s) =_{\text{def}} (sort^*(split(s^+), 0, n))^-$ **where**

$\textbf{inath}^+ = (\textbf{inat}^+\ i: i \le (|S|\ \textbf{div}\ 2)+1),$

$sort^*: ((\textbf{inath}^+ \to \textbf{m}^+) \times (\textbf{inath}^+ \to \textbf{m}^+) \times \textbf{nat} \times \textbf{inat}) \to (\textbf{inat}^+ \to \textbf{m}^+),$
$\textbf{defined}(sort^*(o, e, i, n)) \Rightarrow isext(merge(o, e))\ \Delta\ n = |(merge(o, e))^-|\ \Delta$
$\qquad \#invs((merge(o, e))^-) \le n(n-i)/2,$

$sort^*(o, e, i, n) =_{\text{def}}$
$\quad \textbf{if } i \ge n \textbf{ then } merge(o, e) \textbf{ else } sort^*(transp^*(o, e), i+2, n) \textbf{ fi where}$

$\quad transp^*: ((\textbf{inath}^+ \to \textbf{m}^+) \times (\textbf{inath}^+ \to \textbf{m}^+)) \to$
$\qquad ((\textbf{inath}^+ \to \textbf{m}^+) \times (\textbf{inath}^+ \to \textbf{m}^+)),$
$\quad transp^*(o, e) =_{\text{def}} (e'', d^-(o''))$ **where**

$\qquad ((\textbf{inath}^+ \to \textbf{m}^+)\ o', (\textbf{inath}^+ \to \textbf{m}^+)\ e') = transpo^*(o, e);$
$\qquad ((\textbf{inath}^+ \to \textbf{m}^+)\ o'', (\textbf{inath}^+ \to \textbf{m}^+)\ e'') = transpo^*(d^+(e'), o');$
$\qquad transpo^*: ((\textbf{inath}^+ \to \textbf{m}^+) \times (\textbf{inath}^+ \to \textbf{m}^+)) \to$
$\qquad\quad ((\textbf{inath}^+ \to \textbf{m}^+) \times (\textbf{inath}^+ \to \textbf{m}^+)),$
$\qquad transpo^*(o, e) =_{\text{def}}$
$\qquad\quad ((\textbf{inath}^+\ i:)\ \textbf{m}^+: (mm^+(o(i), e(i))).1,$
$\qquad\quad (\textbf{inath}^+\ i:)\ \textbf{m}^+: (mm^+(o(i), e(i))).2).$

➤ *Representation* in terms of skeletons

The function $transpo^*$ can be folded into

$transpo^*(o, e) =_{\text{def}} MAP2\text{-}2(mm^+, o, e).$

The calls of $d^+$ and $d^-$ within *transp** can be folded with *SHIFTL* and *SHIFTR*:

$transp^*(o, e) =_{def} (e'', SHIFTL(o'', 1, \mathbf{d}))$ **where**
$((\mathbf{inath}^+ \rightarrow \mathbf{m}^+)\ o', (\mathbf{inath}^+ \rightarrow \mathbf{m}^+)\ e') = MAP2\text{-}2(mm^+, o, e);$
$((\mathbf{inath}^+ \rightarrow \mathbf{m}^+)\ o'', (\mathbf{inath}^+ \rightarrow \mathbf{m}^+)\ e'') =$
$MAP2\text{-}2(mm^+, SHIFTR(e', 1, \mathbf{d}), o');$

➤ final polishing

As a last step, we define

$$sort^{**}(o, e, i, n) =_{def} (sort^*(o, e, i, n))^-$$

and calculate a definition of *sort*** which is independent of *sort**:

$sort^{**}(o, e, i, n)$
≡ [ unfold *sort*** ]
$(sort^*(o, e, i, n))^-$
≡ [ unfold *sort** ]
$(\mathbf{if}\ i \geq n\ \mathbf{then}\ merge(o, e)\ \mathbf{else}\ sort^*(transp^*(o, e), i+2, n)\ \mathbf{fi})^-$
≡ [ distributivity of function call over conditional ]
$\mathbf{if}\ i \geq n\ \mathbf{then}\ (merge(o, e))^-\ \mathbf{else}\ (sort^*(transp^*(o, e), i+2, n))^-\ \mathbf{fi}$
≡ [ define *merge** by $merge^*(o, e) =_{def} (merge(o, e))^-$; (4.4-14)
fold *sort*** ]
$(\mathbf{if}\ i \geq n\ \mathbf{then}\ merge^*(o, e)\ \mathbf{else}\ sort^{**}(transp^*(o, e), i+2, n)\ \mathbf{fi}.$

From

$$split^*(s) =_{def} split(s^+)$$

and (4.4-14), independent definitions of *split** and *merge** can be calculated in an analogous way.

Thus as our final result we obtain:

$sort(s) =_{def} sort^{**}(split^*(s), 0, n)$ **where**

$\mathbf{inath}^+ = (\mathbf{inat}^+\ i: i \leq (|S|\ \mathbf{div}\ 2)+1)),$

$sort^{**}: ((\mathbf{inath}^+ \rightarrow \mathbf{m}^+) \times (\mathbf{inath}^+ \rightarrow \mathbf{m}^+) \times \mathbf{nat} \times \mathbf{inat}) \rightarrow (\mathbf{inat} \rightarrow \mathbf{m}),$
$sort^{**}(o, e, i, n) =_{def}$
**if** $i \geq n$
**then** $merge^*(o, e)$
**else** $sort^{**}(e'', SHIFTL(o'', 1, \mathbf{d}), i+2, n)$ **where**

$$((\mathbf{inath}^+ \to \mathbf{m}^+)\, o', (\mathbf{inath}^+ \to \mathbf{m}^+)\, e') = MAP2\text{-}2(mm^+, o, e),$$
$$((\mathbf{inath}^+ \to \mathbf{m}^+)\, o'', (\mathbf{inath}^+ \to \mathbf{m}^+)\, e'') =$$
$$MAP2\text{-}2(mm^+, SHIFTR(e', 1, \mathbf{d}), o')\ \mathbf{fi},$$

$$split^*\colon\ (\mathbf{inat} \to \mathbf{m}) \to ((\mathbf{inath}^+ \to \mathbf{m}^+) \times (\mathbf{inath}^+ \to \mathbf{m}^+)),$$
$$split^*(s) =_{\mathrm{def}} ((\mathbf{inath}^+\ i)\ \mathbf{m}^+\colon \mathbf{if}\ 2i{-}1 > |s|\ \mathbf{then}\ \mathbf{d}\ \mathbf{else}\ s(2i{-}1)\ \mathbf{fi},$$
$$(\mathbf{inath}^+\ i)\ \mathbf{m}^+\colon \mathbf{if}\ 2i > |s|\ \mathbf{then}\ \mathbf{d}\ \mathbf{else}\ s(2i)\ \mathbf{fi}),$$

$$merge^*\colon ((\mathbf{inath}^+ \to \mathbf{m}^+) \times (\mathbf{inath}^+ \to \mathbf{m}^+)) \to (\mathbf{inat} \to \mathbf{m}),$$
$$merge^*(o, e) =_{\mathrm{def}} (\mathbf{inat}\ i)\ \mathbf{m}\colon \mathbf{if}\ \mathbf{even}\ i\ \mathbf{then}\ e(i\ \mathbf{div}\ 2)\ \mathbf{else}\ o((i{+}1)\ \mathbf{div}\ 2)\ \mathbf{fi}.$$

The algorithm we have derived here differs from those in the literature (cf. [1], [6], [10]). The latter algorithms require a number of processors which is equal to the length of the input array and decide on the active processors inside the algorithm by a conditional. Our algorithm needs only half as many processors all of which are active all the time. Whether this difference is sufficient to call our algorithm "new", is left to the reader.

# 5 Conclusions

Apart from being based on ideas outlined in [4], our work presented above has much in common with work done by others, mainly with respect to the use of the functional paradigm and the global transformational development strategy.

The use of functional reasoning, although in different notation, also can be found in e.g. [7] or [13]. As the latter we also prefer to work with arrays defined by functions rather than by data structures.

Our global development strategy is essentially the same as Pepper's [9] or Smith's [11]. Compared to the latter, we assume a user-driven scenario, ignoring questions of mechanizability for the time being. In addition to both, rather than assuming an operational specification to be transformed, we also start from non-operational ones when appropriate. Differently to Pepper, we emphasize the very first part of the development, i.e., the part leading to a functional program composed of skeletons, and do not deal with the problem of implementing these forms which would entail clever scheduling of processors and similar problems.

The work reported on in this paper is to be seen as a first step towards extending the transformational methodology in order to be applicable to the formal derivation of algorithms executable on parallel architectures. Although we restricted ourselves in various respects within this case study (cf. remarks in section 1), we believe that our positive results can be taken as an indication for a success of the underlying approach in general. Similar positive experiences with respect to other aspects of formally deriving prarallel algorithms (cf., e.g., [9], [4]) in other subfields of parallel algorithms back this opinion.

## Acknowledgement

Several people have influenced this paper by constructive criticism and suggestions, most notably E. Boiten, P. Frederiks, M. Geerling, D. Tuijnman, N. Völker, and the unknown reviewers. Their contributions are herewith gratefully acknowledged.

## References

[1] Akl, S.G.: The design and analysis of parallel algorithms. Englewood Cliffs, N.Y.: Prentice-Hall 1989

[2] Broy, M.: A case study in program development: sorting. Institut für Informatik, TU München, TUM-INFO-7831 (1978)

[3] Burstall, R.M., Darlington, J.: A system for developing recursive programs. Journal ACM **24**:1, 44-67 (1977)

[4] Darlington, J., Field, A.J., Harrison, P.G., Harper, D., Jouret, G.K., Kelly, P.J., Sephton, K.M., Sharp, D.W.: Structured parallel functional programming. Imperial College, London, UK, Technical Report 1991

[5] Feather, M.S.: A survey and classification of some program transformation approaches and techniques. In: Meertens, L.G.L.T. (ed.): Program specification and transformation. Amsterdam: North-Holland 1987, pp. 165-196

[6] Goodman, S.E., Hedetniemi, S.T.: Introduction to the design and analysis of algorithms. New York: McGraw-Hill 1977

[7] Meertens, L.G.L.T.: Constructing a calculus of programs. In: Van de Snepscheut, J.L.A. (ed.): Mathematics of program construction. Lecture Notes in Computer Science **375**, Berlin: Springer 1989, pp. 66-90

[8] Partsch, H.A.: Specification and transformation of programs – a formal approach to software development. Berlin: Springer 1990

[9] Pepper, P.: Deductive derivation of parallel programs. Technische Universität Berlin, Fachbereich Informatik, Technical Report 92-23, 1992. Also: this volume

[10] Quinn, M.J.: Designing efficient algorithms for parallel computers. New York: McGraw-Hill 1987

[11] Smith, D.R., Lowry, M.: Algorithm theories and design tactics. In: Van de Snepscheut, J.L.A. (ed.): Mathematics of program construction. Lecture Notes in Computer Science **375**, Berlin: Springer 1989, pp. 379-398

[12] Völker, N.: A formal derivation of odd-even transposition sort. KU Nijmegen, Technical Report 1992

[13] Yang, J.A., Choo, Y.: Design, implementation, and applications of a metalanguage for parallel-program transformations. Dept. of Comp. Sc., Yale University, Technical Report 1990

# 4

# The Use of the Tupling Strategy in the Development of Parallel Programs

Alberto Pettorossi, Enrico Pietropoli
Electronics Department, University of Roma Tor Vergata, 00133 Roma, Italy
adp@iasi.rm.cnr.it

Maurizio Proietti
IASI CNR, Viale Manzoni 30, 00185 Roma, Italy
proietti@iasi.rm.cnr.it

**Abstract**
We present a technique for developing efficient functional programs by using optimal synchronizations among function calls. This technique is based on the application of the tupling strategy and it works for a wide class of functions defined by recursive equations. The derived programs are shown to have minimal redundancy, in the sense that repeated computations of identical recursive calls are never performed and identical subexpressions may be recomputed a number of times which is linearly bounded w.r.t. the depth of the recursion. The derived programs also have minimal spatial synchronization, in the sense that only a minimal number of computations are synchronized together, without increasing the parallel computation time.

## 1 Introduction

When designing and developing algorithms we face, among others, the following problems: how can the required computations be divided into 'small and easy' steps? what functions and data structures should we introduce? how can we improve the efficiency of the preliminary version of the algorithm we have derived?

Those problems are, in general, very difficult to solve. There are, however, some standard techniques which can be applied in a large number of cases to guide the programmer towards satisfactory solutions.

The study and the development of those techniques and similar tools for program construction, is an important research area in Software Engineering. In particular, in the literature there are many program derivation methods which can be applied when programming *in the small*, that is, when constructing small program modules (see, for instance, [26]). They suggest to the programmer the suitable auxiliary functions and/or data structures to be introduced for the derivation of efficient programs.

Many of those methods have been derived by revisiting various algorithm develop-

This work has been partially supported by the "Progetto Finalizzato Sistemi Informatici e Calcolo Parallelo" of the CNR and by MURST 40%, Italy.

ments. Some of them have been sufficiently formalized to allow the construction of powerful automatic systems for algorithm development, like those described in [15, 32]. They are actually used to derive efficient programs for many interesting classes of problems.

Those systems do have a theoretical limitation, in the sense that there will be always some algorithm derivations that they cannot perform no matter how powerful they will be constructed.

In this paper we will look at a particular technique for program development, called the *tupling strategy* [6, 29], and we will study its power in deriving *parallel and synchronized* programs in the case of functional languages. The tupling strategy can be incorporated into automatic program development systems and to some extent this has already been done in the case of logic languages [1].

Before entering into the details of the application of that strategy, let us now briefly recall the basic ideas of the program transformation methodology, according to the so-called 'rules + strategies' approach.

Following that methodology, a program is derived from its specification (or initial program version) by applying some transformation *rules*, which preserve correctness. This derivation process is guided by some metarules, also called *strategies*, which have the objective of producing an efficient final program.

The transformational approach to program development can be described as follows.

> Given the initial program version we first perform some *symbolic evaluation steps*, which usually consist in the application of the so-called *unfolding rule*, and then we look for some properties of the program obtained by those symbolic evaluations. Those properties suggest to us the strategies to be applied (either the tupling strategy or other strategies) and the auxiliary functions to be introduced.
> We then apply some program transformation rules according to the chosen strategy. We also generate the programs for the auxiliary functions, and we express the initial program in terms of those functions. By doing so, we get a new program version which, if some suitable conditions are satisfied, is more efficient than the initial version.

We will be more specific about this transformation process in the following sections where we will also present some examples of its application.

By program transformation it is possible to derive correct and efficient programs without being involved in the invention of complicated loop-invariants. It is also possible to apply the same strategy in the derivations of many different algorithms. This often avoids the repetition of the same correctness proofs during various program developments.

In order to realize the transformational development of parallel programs one should consider both: i) the model of the underlying parallel *machine*, and ii) the *language* used for writing programs. The machine model has to be considered because it determines the relevant notions of synchronization among processes and the complexity measures for evaluating program efficiency. Also the language used for writing programs should be considered, because as we will see, it plays a crucial role in the development of *syntactic* techniques for transforming programs.

For the construction of program transformation systems it is important to take into account, together with the machine model and the language, the *metalanguage* used for specifying the program transformations themselves. Little attention has been devoted in the literature to this issue, although some ideas have been presented in [11, 12, 15] and some techniques have been developed in the area of Theorem Proving, like for instance, the *tacticals* of the Edinburgh LCF [19] and the various methods for 'plan description' in Artificial Intelligence. We will not address this issue here.

The paper is structured as follows. In Section 2 we introduce the computational model for the parallel and synchronized execution of functional programs. In Section 3 we present the notion of a *symbolic evaluation step* for those programs and we show how to apply the tupling strategy if the programs produced by the symbolic evaluation steps satisfy some given properties. By using that strategy we can automatically develop parallel programs with optimal synchronizations among function calls.

In Section 4 we will study the power of the tupling strategy, that is, its ability to derive efficient parallel algorithms for some classes of program specifications. We consider two particular forms of synchronization, namely temporal and spatial synchronization, and we analyze their properties with respect to both the avoidance of repeated computations and the need for computing processes.

In Section 5 we present a procedure which by making use of the tupling strategy, synthesizes optimal synchronized parallel programs for the evaluation of a given class of functions defined by recursive equations. Finally, in Section 6 we present some examples of the application of that procedure.

# 2 A Computational Model for Functional Languages

In this section we will present our computational model for the parallel execution of programs in the case of typed functional languages, such as ML or Miranda. With reference to that model we will introduce the notion of parallel and synchronized evaluation of functional expressions, and we will formalize the program transformation process and the application of the tupling strategy, which is our interest here.

We start off by considering the following pairwise disjoint sets:

i) *Typed Variables*: x, y, z, ... ∈ Vars.

ii) *Typed Function Symbols* (with arities): f, g, ... ∈ Fncts.
Each function symbol denotes a typed function, which is defined by recursive equations (see below).

iii) *Typed Constructors* (with arities): constr ∈ Constr.

| We have, for instance: | | |
|---|---|---|
| | true, false | for boolean values |
| | 0, s(_) | for natural numbers |
| | nil, cons(_,_) | for lists |
| | empty, tip(_), tree(_,_) | for binary trees |
| | <_,...,_> | for tuples with $n \geq 1$ components. |

We will introduce some more typed constructors as the need for different types (or data structures) arises. For brevity, we will also write k instead of $s^k(0)$, where $s^0(x)$ is x and $s^{k+1}(x)$ is $s(s^k(x))$. We will also write n+k instead of $s^k(n)$.
The unary tuple constructor is the identity function, so that <x> is identical to x.

iv) *Typed Basic Operators* (with arities): bop ∈ BOp.
Particular basic operators are the *destructors* which correspond to constructors [23]. For instance, in correspondence to tree(_,_) we have the two destructors left(_) and right(_), such that left(tree(L,R)) = L and right(tree(L,R)) = R, for any non-empty binary tree t.
For any i = 1,...,n we have the destructor πi, that is, the i-th projection function, which is defined as follows: $\pi i(\langle a_1,\ldots,a_i,\ldots,a_n\rangle) = a_i$. Obviously, $\pi 1\langle x\rangle = x$.
Each basic operator denotes a typed function which is specified by the language semantics, *not* by recursive equations (see below).

From the above sets we have the following derived sets, where n≥0 and k≥1:

i) *Elementary Expressions*: ee ∈ EExpr ee ::= x | $\text{constr}(ee_1,\ldots,ee_n)$
We will also consider *Constructor Expressions*, CExpr for short, which are Elementary Expressions made out of constructors only. Constructor Expressions are ranged over by the variable ce, possibly with subscripts.

ii) *Expressions*: e ∈ Expr e ::= ee | $\text{bop}(e_1,\ldots,e_n)$ | $f(e_1,\ldots,e_n)$
| **if** $e_0$ **then** $e_1$ **else** $e_2$
| $e_0$ **where** $\langle x_1,\ldots,x_k\rangle = e$

The expression **if** $e_0$ **then** $e_1$ **else** $e_2$, also written if-then-else($e_0,e_1,e_2$), is called a *conditional*. The subexpressions $e_0$, $e_1$, and $e_2$ are called *condition*, *left arm*, and *right arm*, respectively. The expression $e_0$ **where** $\langle x_1,\ldots,x_k\rangle = e$ is called a *where-clause* and $\langle x_1,\ldots,x_k\rangle = e$ is called its *definition*. $x_1,\ldots,x_k$ are the *bound* (or *defined*) *variables* of the where-clause. This notion of bound variables is consistent with the λ-calculus, because e **where** x=b can be viewed as (λx.e) b and x is a bound variable in λx.e.
We assume that all expressions are well typed and well formed, that is, the types and the number of the arguments agree with the arities.

In what follows we also need the following notions.
An *operator* is an element of Constr ∪ BOp ∪ {if-then-else}. The expression $e_1$ is an *instance* of the expression $e_2$ iff there exists a matching substitution σ so that $e_1=\sigma(e_2)$. For example, we have that f(tree(empty,tip(y))) is an instance of f(tree(empty,x)) with the substitution σ = {x=tip(y)}.
A *context expression* C[...] is an expression C with a missing subexpression. For instance, if the context expression C[...] is f(tree(...,x)) then C[tip(z)] is f(tree(tip(z),x)). We also need the usual notions of *free* and *bound variables* of a given expression [4].

iii) *Recursive Equations*: req ∈ REq req ::= $f(ee_1,\ldots,ee_n) = e$
They are also called *equations*, for short. We assume that in a recursive equation any variable in the r.h.s., which is not a bound variable, occurs also in the l.h.s.

iv) A *recursive equations program* (also called *program*, for short) is a set of recursive equations. Given a program P, if one of its recursive equations has its left hand side of the form f(...) we say that P *defines* the function f.
We assume that the recursive equations of a given program P are: i) *mutually exclusive*, that is, the l.h.s.'s of any two equations do not have a common instance, and ii) *exhaustive*, in the sense that for each function f (of arity n) defined by P and for each n-tuple of values $v_1,\ldots,v_n$ in the domain of f we have that $f(v_1,\ldots,v_n)$ is an instance of the l.h.s. of an equation of P.

We can always write a program in a form, which is called *tree form*, obtained as follows:

- we eliminate all where-clauses (and we can do so by writing: $e_0[x_1/\pi 1(e),\ldots, x_k/\pi k(e)]$, instead of $e_0$ **where** $\langle x_1,\ldots,x_k\rangle = e$), and
- for any given function symbol f we replace all recursive equations whose l.h.s. is of the form f(...) by a single equation of the form: $f(x_1,\ldots,x_n) = e$, where $x_1,\ldots,x_n$ are distinct variables.
  We can always do so by using: i) conditionals for distinguishing the different cases, and ii) auxiliary basic operators (like the equality predicate, suitable destructors corresponding to given constructors, etc.).

Obviously, the tree form of a program is *not* uniquely determined. However, the results we will present in the paper do not depend on the particular tree form we may consider.
It can be shown that the transformation of a program in its tree form preserves the semantics of a program as we will define it below.

**Example 1** Let Nat be the set of natural numbers. Let us also consider the auxiliary destructor operator p(_) which denotes the predecessor function on natural numbers and corresponds to the constructor s(_). The tree form of the following program:
1.1 f(0) = a
1.2 f(n+1) = b(n, f(n))
which defines the function f: Nat → Nat, is: f(n) = **if** n=0 **then** a **else** b(p(n), f(p(n))).
The tree form of the following program:
1.3 h(0) =1
1.4 h(2) = 1
1.5 h(n+3) = h(m) **where** m = n+2 for $n\geq 0$
which defines the function h: (Nat – {s(0)}) → Nat, is:

h(n) = **if** n=0 **then** 1 **else if** n=2 **then** 1 **else** h(p(n)).

Notice that the expression h(1) is *not* well typed and it will never arise during any computation of h. ■

The semantics of a program can be given in an operational or denotational manner as indicated in [10, 27, 34]. We now define the *operational semantics* S[e] of any expression e without free variables. In order to do so we assume that we are given a program, say P, in tree form and an interpretation, say I, of the basic operators in BOp. For any $n\geq 0$ and for any basic operator bop of arity n, the value of I(bop), also denoted by $bop_I$, is a function from $CExpr^n$ to CExpr.

Thus, to be more precise, the semantics S[e] of the expression e should be written as $S_{P,I}[e]$, because it depends on the program P and the interpretation I.

Given an expression e, S[e] is operationally defined by a *parallel graph rewriting* procedure, called the *evaluation* (or *computation*) of e, which returns, if it terminates, a graph whose nodes are labelled by constructor expressions only (see, for instance, [4]), as we now specify.

Given the expression e we first consider the *computation graph* G[e] constructed as follows.

**Construction of the computation graph** G[e].

i) Given the expression e we take the usual tree representation, called T[e], of e according to the arities of the operators and function symbols. In particular:
   - the tree corresponding to any expression of the form: **if** $e_0$ **then** $e_1$ **else** $e_2$ is the one for the operator if-then-else with the three arguments $e_0$, $e_1$, and $e_2$, and
   - the tree corresponding to any expression of the form: $e_0$ **where** $<x_1,\ldots,x_k> = e$ is the one for $e_0$ where each leaf $x_i$ has been replaced by a copy of the tree corresponding to $\pi i(e)$, for $i=1,\ldots,k$.

ii) we identify *some* (none or more) nodes corresponding to syntactically identical subexpressions. We assume that in the graph of $e_0$ **where** $<x_1,\ldots,x_k> = e$ we have only one node corresponding to the expression e (see Fig. 1). ■

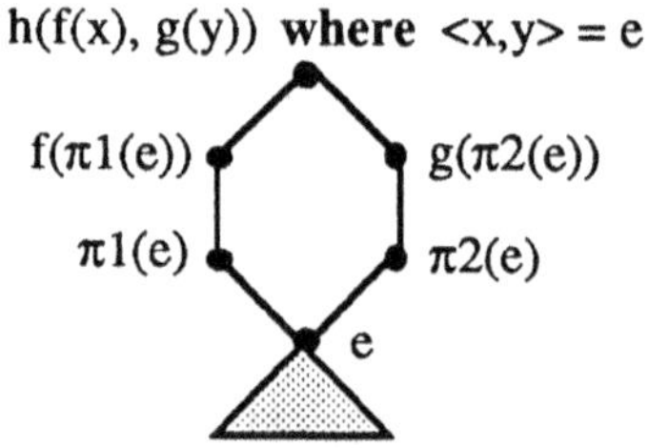


Figure 1: The graph corresponding to the expression: h(f(x), g(y)) **where** <x,y> = e, that is, h(f(π1(e)),g(π2(e))).

The construction of G[e] is *nondeterministic* and it may be very efficiently implemented, because it does not require an exhaustive search over the tree T[e] when looking for syntactically identical subexpressions. Notice that for any given expression e, the semantics S[e] does not depend on the way the nondeterminism of this construction is actually resolved.

We say that the graph G[e] has *label* e. We also assume that every node in G[e] is labelled by the corresponding subexpression of e.

In the figures, for simplicity, we often indicate for each node only the topmost symbol of its label. For instance, G[+(s(s(0)), s(0))] is depicted in one of the following two ways:

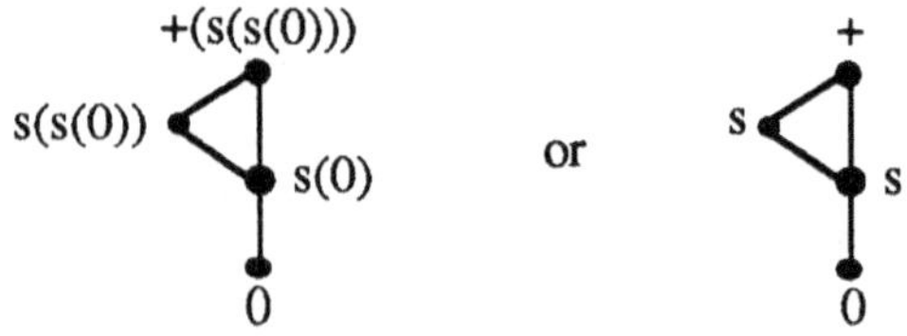


To compute S[e] we apply to the graph G[e] a *maximal* sequence of *parallel*

*rewriting steps*. Each step of that sequence consists in:

i) choosing in the graph in hand one or more disjoint subgraphs, each of which is linked to the remaining graph only through the topmost node, and

ii) rewriting each of those subgraphs in parallel. The rewriting of a chosen subgraph is realized by applying one of the rules R1, R2, or R3 presented below, while complying with the metarule MR, also presented below.

The sequence is maximal in the sense that it cannot be made longer by a further rewriting step. The resulting graph (and the corresponding label), if any, is the operational semantics S[e] of the given expression e.

Let us now present the rules R1, R2, and R3.

R1. For any expression bop($ce_1,\dots,ce_n$), where $ce_1,\dots,ce_n$ are constructor expressions:

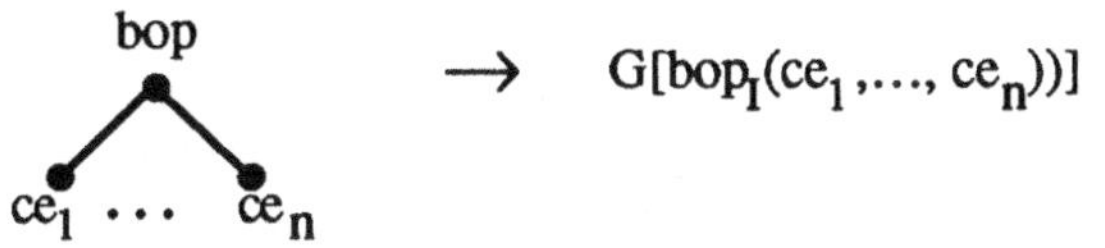


R2. For any constructor expression c, and expressions $e_1$ and $e_2$:

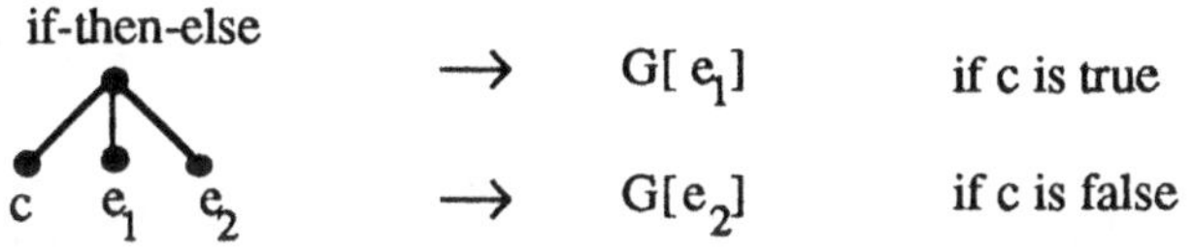


R3. (*Unfolding Rule*) If the equation f($ee_1,\dots,ee_n$) = e occurs in the program P and the constructor expressions $ce_1,\dots,ce_n$ are instances of $ee_1,\dots,ee_n$ via a substitution σ:

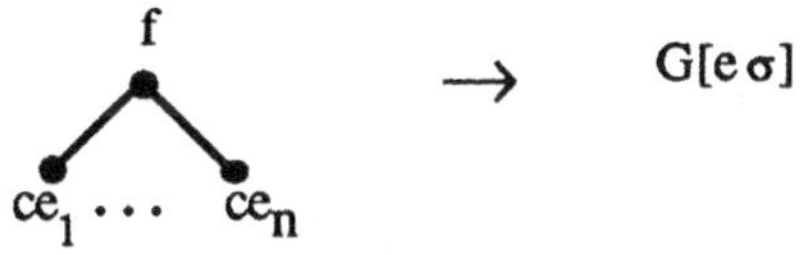


In this case we say that rule R3 is applied to the function symbol f. ■

**Remarks** i) We assume that rules R1, R2, and R3 can also be applied when some of the nodes of the graphs in the left hand sides have been identified because they correspond to the same subexpression.

ii) A subgraph can be transformed into one which is linked to the remaining graph only through the topmost node by duplicating shared nodes as depicted in Fig. 2. It is *not* allowed to duplicate the nodes corresponding to definitions of where-clauses. (Otherwise we break the rules for the construction of our graphs.)

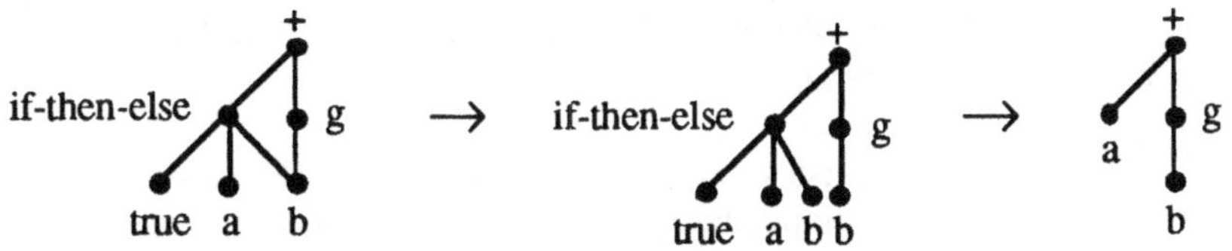


Figure 2: Duplication of the shared node b and rewriting of the if-then-else. ■

In order to present the metarule MR we need the following notion. Consider an expression e, its corresponding graph G[e], and a particular *occurrence* f* of the function symbol f in G[e]. The set of *controlling conditions of* f* *in* G[e], denoted by conds(f*,G[e]), is the following set of occurrences of subexpressions of e:

$$\text{conds}(f^*,G[e]) =_{def} \{ e_0 \mid \textbf{if } e_0 \textbf{ then } e_1 \textbf{ else } e_2 \text{ occurs in } G[e] \text{ and } f^* \text{ occurs in } G[e_1] \text{ or } G[e_2] \}.$$

Thus, in order to find conds(f*,G[e]) it is enough to 'move up' from f* to the topmost symbol in G[e] and collect the conditions of the if-then-else's where f* occurs *within* either the left or the right arm.

Since for any given expression e the tree T[e] is a particular graph of computation G[e], where there are no identifications of nodes, we have that: conds(f*,T[e]) $\subseteq$ conds(f*,G[e]).

Notice also that conds(f*,G[e]) is uniquely determined by f* and G[e], not by f* and e. Indeed, let us consider the graph $G_1$ for the expression e = +(**if** p **then** f(0) **else** 0, **if** q **then** f(0) **else** 0) which is constructed by identifying the two subgraphs for f(0). We have: conds(f*,$G_1$) = {p,q}, where f* is the only occurrence of f in $G_1$. On the other hand, if we construct the graph $G_1$ for the above expression e by maintaining two distinct copies for f(0), we get either conds(f*,$G_1$) = {p} or conds(f*,$G_1$) = {q}, according to which occurrence f* in $G_1$ we choose.

The metarule MR is as follows.

MR. If during a parallel rewriting step we want to apply rule R3 in a subgraph of the graph in hand, say G[$e_1$], possibly in parallel with other instances of the rules R1, R2, and R3, then we should apply rule R3 at least to *either*

i) the leftmost innermost occurrence, say f*, of a function symbol f occurring inside an outermost conditional, such that rules R1 and R2 *cannot* be applied to any expression in conds(f*,G[$e_1$]) *or*

ii) a function symbol occurring outside all conditionals of G[$e_1$].

(The notions of *leftmost*, *inside*, and *outside* are the usual ones w.r.t. the tree representation T[e] of a given expression e.) ■

Notice that the form of rule R3 implies that in *both* cases, MR.i) and MR.ii), the occurrence, say f*, of the function symbol of the l.h.s. is an *innermost* one (that is, no other function symbol occurs in any argument of f*) and no further applications of the rules R1 and R2 can be made inside the arguments of f*. In case MR.i) f* must be inside the condition of a conditional.

Our operational semantics is a parallel variant of the *call-by-value semantics*, as usually considered in the literature. Son-nodes are rewritten before the corresponding father-nodes. It is clear that the metarule MR allows for the parallel rewriting of the graph in hand in many different ways.

The graph rewriting process we have presented above, determines the model of parallel and synchronized computation which we now describe.

Given an expression e to be evaluated, we first associate a *rewriting process* (or simply, a *process*) with each node of G[e] whose label is not a constructor expression. These processes concurrently rewrite the graph to which they refer, following the rules R1, R2, R3, and the metarule MR. (Recall that there are no rules for rewriting nodes whose label is a constructor expression.) As already mentioned, disjoint portions of the graphs may be rewritten in parallel.

While the rewriting of the graph progresses, new graphs are generated and new processes are associated with each node whose label is not a constructor expression. Expressions (or subexpressions) can be rewritten in parallel if their graphs do not share any node. If they do share some nodes, before allowing for their parallel rewriting, we need to duplicate the shared portions of the graphs.

These duplications may, in general, generate an exponential number of nodes, and since it is unreasonable to have at our disposal an exponential number of processes, we assume the existence of an allocation procedure for assigning the available processes to the nodes of the graph in hand. This allocation procedure determines the portions of the graph where the rewritings can take place, because of the presence of processes at the corresponding nodes.

*Synchronizations* occur when we rewrite nodes associated with basic operators or function symbols (see rules R1 and R3), in the sense that in our call-by-value semantics the evaluation of a node can only be performed when all son-nodes have been evaluated to constructor expressions. In particular, any projection operator $\pi i$ forces a synchronization among the various components of its argument. For instance, in the case of the expression h(f(x), g(y)) **where** <x,y> = e (see Fig. 1) the evaluation of e, and thus, the ones of x and y, must be completed before the one of f *or* the one of g may begin. Thus, where-clauses of the form: expr[$e_1,\ldots,e_n$] **where** <$e_1,\ldots,e_n$> = e can be used for synchronizing the computations of the expressions $e_1,\ldots,e_n$ by making them into the tuple <$e_1,\ldots,e_n$>.

During program development we face the problem of avoiding the repeated evaluation of common subexpressions for increasing the efficiency of our programs. This can be done by deriving some suitable synchronizations among computations, that is, suitable where-clauses and tuples. As we will see below, these synchronizations can be determined in an automatic way by program transformation through the application of the tupling strategy. Unfortunately, the use of this strategy may sometimes reduce parallelism, because it generates shared nodes in the computation graph. However, it is possible to achieve a higher efficiency, because the desired computations can often be performed within the same total computation time, using a linear number of processes, instead of an exponential one.

A different form of synchronization among basic operators or function calls, takes place at a shared son-node. For instance, if f(x,e) and g(e,y) (see Fig. 3) need to be evaluated and there exists only one node with label e then the evaluation of e must be completed before the beginning of the evaluation of both f and g.

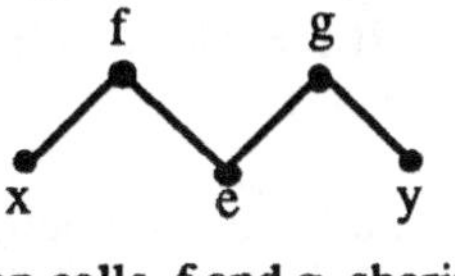


Figure 3: Two function calls, f and g, sharing the same argument e.

Related approaches to the parallel execution of functional languages have been proposed in the literature. In particular, we would like to mention: i) the Alice proposal

[13], where each function evaluation is represented by a so-called *packet*, which basically corresponds to a node of our graphs, and ii) the approach described in [31], where a *process* is assumed to be a triple of a *unique name* (or address), a *local memory* for exchanging information among processes, and an *expression* which should be evaluated by that process.

Now we will show that given a program P, the parallel rewriting process defined by rules R1, R2, R3, and MR computes the least fixpoint of P.

Without loss of generality, we may assume that the program P consists of the equation $f(x_1,\dots,x_n) = e$ only, written in tree form. The case of k equations for k>1, can be derived from the case where k=1, in a straightforward way by considering k-tuples of functions, instead of one function only.

We assume that a data type is a domain of the form: $D = \{\bot\} \cup S$, where S is a given set of constructor expressions, like nil, true, false, 0, s(0), etc. In this case D is also denoted by $S_\bot$.

Any domain D is considered to be a *flat domain*, that is, the ordering $\le_D$ on D is defined as follows: $a \le b$ iff ($a = \bot$ or $a = b$). We will write $\le$, instead of $\le_D$, when the domain D is understood from the context.

We say that $\langle x_1,\dots,x_n\rangle \le \langle y_1,\dots,y_n\rangle$ iff $x_i \le y_i$ for each i = 1, ..., n.

A function g of arity n from $D_1\times\dots\times D_n$ to D is said to be *strict* iff if one of its arguments is $\bot$ then the value of g is $\bot$.

An interpretation $b_I$ of a basic operator b in BOp of arity n is assumed to satisfy the following conditions: i) it is a *strict function* from $D_1\times\dots\times D_n$ to D for some suitable domains $D_1,\dots, D_n$, D, and ii) if $x_1\neq\bot$ and... and $x_n\neq\bot$ then $b_I(x_1,\dots,x_n) \neq\bot$.

As a consequence, the projection function $\pi i$ returns $\bot$ if *any* of the components of its argument is $\bot$.

The conditional is assumed to be *sequential*, that is, the following conditions hold: i) **if** $\bot$ **then** $e_1$ **else** $e_2 = \bot$, ii) **if** true **then** $e_1$ **else** $e_2 = e_1$, and iii) **if** false **then** $e_1$ **else** $e_2 = e_2$.

We also assume that the evaluations always agree with the types of the expressions to be evaluated. For instance, we assume that we are never asked to evaluate (or rewrite) expressions like +(1,true).

A function $g_I$ of arity n from $D_1\times\dots\times D_n$ to D is said to be *monotonic* iff for all $x_1,\dots,x_n$, $y_1,\dots,y_n$ ranging over $D_1\times\dots\times D_n$ if $\langle x_1,\dots,x_n\rangle \le \langle y_1,\dots,y_n\rangle$ then $g_I(x_1,\dots,x_n) \le g_I(y_1,\dots,y_n)$. It is easy to show that for each basic operator $b \in$ BOp, $b_I$ is monotonic.

It can also be shown that each program P defines a continuous functional which is denoted by $\tau_P$, or simply by $\tau$ when P is understood from the context [25].

For instance, given P = $\{f(n) =$ **if** $n=0$ **then** 1 **else** $n \times f(n-1)\}$, we have that: $\tau_P = \{\lambda fn.$**if** $n=0$ **then** 1 **else** $n \times f(n-1)\}$.

Given a program $P = \{f(x_1,\dots,x_n) = e\}$, we define the Kleene sequence of P to be: $<\Omega, \tau_P(\Omega), \tau_P{}^2(\Omega), \dots>$, where $\Omega$ is the undefined function, that is, $\lambda x_1\dots x_n.\bot$, and $\tau_P{}^{i+1}(\Omega)$ is obtained from $\tau_P{}^i(\Omega)$ by a simultaneous replacement of *all* $\Omega$'s by $\tau_P(\Omega)$.

**Theorem 1** A program $P = \{f(x_1,\ldots,x_n) = e\}$, determines the (unique) least fixpoint $f_{fix}$, which is the limit of the Kleene sequence: $<\Omega, \tau_P(\Omega), \tau_P{}^2(\Omega), \ldots>$.
**Proof** [25]. ■

**Theorem 2** Given a program $P = \{f(x_1,\ldots,x_n) = e\}$, if its least fixpoint $f_{fix}$ is strict then $f_{fix}$ is computed by a parallel rewriting process according to the rules R1, R2, R3, and the metarule MR.
**Proof** Let the least fixpoint $f_{fix}$ be a function from $D_1\times\ldots\times D_n$ to D. Let us consider a generic computation sequence $\gamma = <t_0, t_1,\ldots, t_k,\ldots>$ starting from the expression $t_0 = f(c_1,\ldots,c_n)$, where $c_1,\ldots,c_n$ are some chosen elements in the domains $D_1,\ldots,D_n$. Let $\gamma_k$ denote the subsequence $<t_0, t_1,\ldots, t_k>$ of $\gamma$. Without loss of generality, we assume that the transition from $t_k$ to $t_{k+1}$, for any $k\geq 0$, is made by the application of exactly one of the rules R1, R2, and R3, according to the metarule MR.

Let us also assume that any occurrence of f in $t_k$, for $k\geq 0$, is indexed by a natural number. The symbol f in $t_0$ is indexed by 0. These indexes change along the computation sequence as follows:

for $k\geq 0$ each occurrence of f in $t_{k+1}$ which is also an occurrence in $t_k$ (in the sense that it corresponds in the rewriting process to an occurrence in $t_k$), maintains the index it has in $t_k$, while all occurrences of f which are generated by substituting $e\sigma$ for the subexpression $f(x_1,\ldots,x_n)\sigma$ in $t_k$ by rule R3, have index r+1 iff the occurrence of f in $f(x_1,\ldots,x_n)\sigma$ has index r.

Let us consider the function $V: Expr_\perp \rightarrow Cexpr_\perp$, defined as follows:

i) $V(\perp) = \perp$
ii) $V(ce) = ce$ for any $ce \in$ CExpr
iii) $V(g(e_1,\ldots,e_n)) = g_I(V(e_1),\ldots,V(e_n))$ for any n and for any g in BOp of arity n
iv) $V(f(c_1,\ldots,c_n)) = \perp$
v) $V($**if** $e_0$ **then** $e_1$ **else** $e_2) =$ if $V(e_0) =$ true then $V(e_1)$ else if $V(e_0) =$ false then $V(e_2)$ else $\perp$.

In order to prove the theorem we need to show that:
1) for any $k\geq 0$ there exists m such that $V(t_k) \leq \tau^m(\Omega)(c_1,\ldots,c_n)$, and
2) for any $m\geq 0$ there exists k such that $\tau^m(\Omega)(c_1,\ldots,c_n) \leq V(t_k)$.

To show point 1) it is enough to take m larger than all indexes of the f's occurring in the finite subsequence $\gamma_k$. Indeed, for $j=0,\ldots,k$ $V(t_j)$ can be obtained from the expression of $\tau^m(\Omega)(x_1,\ldots,x_n)$ by: i) replacing $(x_1,\ldots,x_n)$ by $(c_1,\ldots,c_n)$, ii) possibly replacing some subexpressions by $\perp$ (recall rule iv), and iii) evaluating some basic operators and conditional expressions.

Point 2) is obvious if $\tau^m(\Omega)(c_1,\ldots,c_n) = \perp$. If $\tau^m(\Omega)(c_1,\ldots,c_n) = a \neq \perp$ then point 2) can be derived from the validity of the following Property ($\alpha$): given the computation sequence $\gamma = <t_0, t_1,\ldots, t_k,\ldots>$ we have that $V(\gamma) = <V(t_0), V(t_1),\ldots, V(t_k),\ldots>$ is *either* infinite and equal to $<\perp, \perp,\ldots, \perp,\ldots>$ *or* it is finite and equal to $<\perp, \perp,\ldots, \perp, a_1,\ldots, a_h>$ where $h\geq 1$ and $a_1 =\ldots= a_h = a \neq \perp$.

We leave to the reader to show that by using Property ($\alpha$) and the strictness of $f_{fix}$, we have: if $\tau^m(\Omega)(c_1,\ldots,c_n) = a$ then $V(\gamma)$ is finite and its last element is a. ■

Our framework for the computation of $f_{fix}$ extends the one of [25], because in [25] between any two unfoldings of the recursive definitions, that is, any two uses of rule R3, one should perform all possible simplification steps corresponding to the various operators, while in our case we have only to comply with the metarule MR and the rewritings can be done in parallel.

In our presentation we did not consider the case of *higher-order functions* and *lazy functions*. A motivation behind our choice is the fact that the implementation 'via parallel rewriting' of functional languages with higher-order functions and laziness, is a hard problem. Some solutions for shared memory multiprocessing are presented in [2, 17, 18, 24]. Laziness can be realized by delayed computations which are often stored in heap structures, where destructive operations take place. The rewritings of the nodes which correspond to delayed computations, are controlled by *lock bits* which are used for enforcing mutual exclusion.

# 3 Synchronization via Tupling

We will now consider the tupling strategy as a technique for synchronizing various function calls in the parallel model of evaluation we have described above. The need for synchronization arises because, as we will show, parallelism alone is not sufficient for an efficient execution of recursive functional programs. Indeed, for these computations it may be necessary an exponential number of processes if synchronization is not used.

Let us consider non-linear recursive programs which are instances of the following equation (or program) P: f(x) = **if** p(x) **then** a(x) **else** b(c(x), f(*l*(x)), f(r(x))).

We assume that the interpretation I of the operators occurring in P is given by the following strict functions: for some given domains X and Y, $p_I$: X → {true, false}$_\perp$, $a_I$: X → Y, $b_I$: X×Y×Y → Y, $c_I$: X → X, $l_I$: X → X, and $r_I$: X → X. In these hypotheses, the least fixpoint solution $f_{fix}$ of the program P is a strict function from X to Y.

We also assume that every call-by-value sequential evalution of f(x) terminates for any x≠⊥, that is, given the function Eval: (X → {true,false}$_\perp$) × X → {true,false}$_\perp$ defined as follows:

Eval($h_I$, x) $=_{def}$ ($h_I$(x) = true) or (Eval($h_I$, $l_I$(x)) and Eval($h_I$, $r_I$(x))),
we have that Eval($p_I$, x) = true, for any x≠⊥.

We now introduce an *abstract model* of computation for the programs described by the above equation P. In this model we assume that $p_I$(x) is always false, and thus, we focus our attention on the recursive structure of the function f(x) and we do not take into account the termination issue. We will indicate below the relationship between this abstract model and the one presented in the previous section by the rules R1, R2, R3 and MR.

**Definition 3** Let us consider the equation P, an interpretation I of its basic operators, and its least fixpoint solution $f_{fix}$: X → Y. The corresponding *symbolic tree of recursive calls* (or *s-tree*, for short) is a directed infinite tree whose nodes are labelled by function calls. It is constructed as follows.

i) The root node, called *initial node*, is labelled by f(x), where x is a variable ranging over X.
ii) For any node p labelled by f(e) for some expression e, there exist two *son-nodes*: the *left-son* $p_l$ with label f(*l*(e)) and the *right-son* $p_r$ with label f(r(e)), and there exist

two directed arcs: $<p,p_l>$ and $<p,p_r>$. We will often associate the label $l$ (or r) with the arc $<p,p_l>$ (or $<p,p_r>$) (see Fig. 4). ■

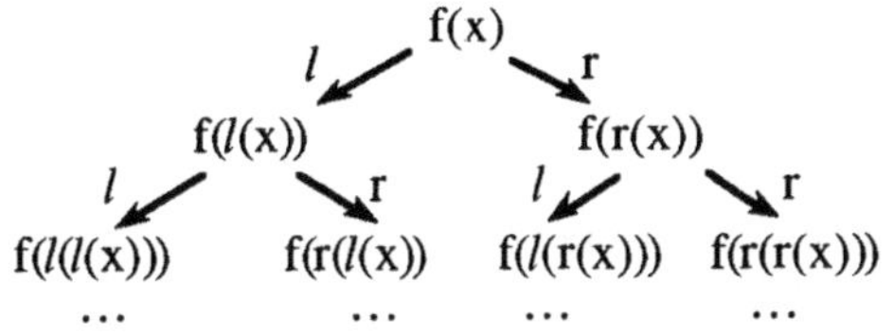


Figure 4: The symbolic tree of recursive calls for the equation:
f(x) = **if** p(x) **then** a(x) **else** b(c(x), f($l$(x)), f(r(x))).

For nodes of s-trees, we also make use of the usual relations of *father*, *ancestor*, and *descendant* nodes. For simplicity, in the sequel we will often say 'the node f(e)', instead of 'the node with label f(e)'. The generation of the two son-nodes of a given node corresponds to a symbolic evaluation step (or unfolding step) of that node. Thus, s-trees may be viewed as an abstract model of computation for the program P.

We can identify each node of an s-tree with a word, called the *associated word*, in the monoid $\Sigma^* = \{\varepsilon\} \cup \Sigma \cup \Sigma^2 \cup \ldots$, generated by $\Sigma = \{l, r\}$, as we now specify: i) we identify the initial node with the empty word $\varepsilon$, and ii) if a node, say p, is identified with the word w then the left-son of p is identified with the word $wl$ and the right-son of p is identified with wr. Thus, each node is identified with the word which indicates the downward path from the root of the s-tree to that node. The above identification establishes a bijection between the set of nodes of any s-tree and $\Sigma^*$.

We say that a node whose associated word is $w = s_1 s_2 \ldots s_n$, for some $s_1, s_2, \ldots, s_n$ in $\Sigma$, is *at level* n. We define the *length* L(w) of a word w in $\Sigma^*$ to be the value n such that $w \in \Sigma^n$. Thus, a node is at level n iff the length of its associated word is n.

**Definition 4** Given the equation P, an interpretation I of its basic operators, and a word $u = s_1 s_2 \ldots s_n$ in $\Sigma^n$ for some $n \geq 0$, the expression $s_{nI}(\ldots(s_{2I}(s_{1I}(x)))\ldots)$ will also be denoted by $u_I(x)$. ■

**Definition 5** Given a symbolic tree of recursive calls the corresponding *symbolic graph of recursive calls* (or *s-graph*, for short) is obtained by identifying any two nodes p and q of the symbolic tree iff $\forall x \in X.\ u_I(x) = v_I(x)$, where u and v are the words associated with p and q, respectively. ■

Notice that in an s-graph we may have more than one arc leading to the same node, and we may have cycles. The existence of cycles does not contradict our assumption that f(x) terminates for any $x \neq \perp$, because when computing f(x) we have to take into account also the value of the predicate $p_I$.

Multiple occurrences of identical calls of the function f in a symbolic tree of recursive calls for the program P and the interpretation I, can be represented by a set of equations between words in $\Sigma^*$. For instance, the fact that $f(l_I(l_I(v))) = f(r_I(v))$ for any v in X, can be represented by the equation: $ll = r$.

Thus, any given s-graph has an associated set $E^c$ of equations defined as follows:

$E^c = \{u = v \mid u, v \in \Sigma^*$ and $\forall x \in X.\ u_I(x) = v_I(x)\}$. $E^c$ is closed under reflexivity, symmetry, and transitivity. It is also closed under left and right congruence, that is, *if* $a \in \Sigma$ and $s = t \in E^c$ *then* $\{as = at, sa = ta\} \subseteq E^c$.

We can identify each node of a given s-graph with an element of the set $\Sigma^*/E^c$, that is, an $E^c$-equivalence class of words in $\Sigma^*$, called the *associated equivalence class*, according to this rule: *if* f(u(x)) is the label of the node p for some word u in $\Sigma^*$, *then* u is in the equivalence class of words, denoted by [u], identified with p.

This identification establishes in any s-graph a bijection between the set of nodes and $\Sigma^*/E^c$.

In what follows we will find useful to associate with any s-graph G a monoid, say M(G), whose carrier is $\Sigma^*/E^c$. The neutral element of M(G) is $[\varepsilon]$, and the concatenation in M(G) is defined as follows: $[u] \cdot [v] =_{def} [u\,v]$. If $E^c = \varnothing$ we say that the monoid M(G) is *free*, and in this case the s-graph is equal to the s-tree.

As it was the case for s-trees, also s-graphs provide an abstract model of computation for the program P. The relationship between this abstract model and the one we have introduced in the previous section, is as follows. Suppose that we are given an interpretation I for the basic operators occurring in the program P and we want to compute the value of f(v) for some v in X. If in the labels of the s-graph we replace x by v and we do not consider the nodes which are descendants of any node [u] with label f(u(v)) for which $p_I(u_I(v))$ = true, then we get a finite labelled graph, say G, which represents the recursive calls of f to be evaluated during the computation of f(v).

Since the labels of the nodes in G are all distinct, no repeated evaluations of identical recursive calls are performed.

Let us now assume that there exists a constant $K > 0$ such that for all values $v_1$, $v_2$, and $v_3$ in $X \times Y \times Y$ the number of processes needed for the evaluation of $b_I(c_I(v_1), v_2, v_3)$ is bounded by K. Then the total number of processes necessary for the evaluation of f(v) according to the model of computation presented in the previous section, is proportional to the number of nodes in the finite graph G. Indeed, we need one process for each node of G and we also need at most K processes to compute the function call labelling the father-node given the values of the function calls labelling the son-nodes.

The existence of the constant K for the evaluation of $b_I(c_I(v_1), v_2, v_3)$ may not be always satisfactory. However, this assumption can be considered as a first step towards a more detailed analysis of the computational performances of our parallel algorithms.

The relationship we have established between the two models of computation, makes it very important the construction of the s-graphs from s-trees. By this construction, infact, we identify nodes of an s-tree and we may reduce by an exponential amount the number of processes needed for the evaluation of f(v) for any given v in X. The following Example 2 will illustrate this point.

The identification of the nodes of an s-tree can be considered as realizing suitable *synchronizations* among the processes which compute the function calls of the nodes which have been identified. Obviously, this synchronization does not increase the parallel time complexity, because the length of the longest path from the initial node to a leaf, is not increased.

There exists another form of synchronization which can be imposed on the nodes of s-graphs without increasing the parallel time complexity. That synchronization is realized by the application of a program transformation strategy, called tupling strategy [30], which is based on the discovery of suitable properties of the s-graph in hand. If those properties do hold, one can generate from the given program an equivalent one which is linear recursive. Then (see Example 2 below) if we assume the existence of the above mentioned constant K for the evaluation of $b_I(c_I(v_1),v_2,v_3)$ for all $v_1$, $v_2$, and $v_3$, the number of processes necessary for the evaluation of f(v) is *linear* w.r.t. the depth of the recursion.

Moreover, as shown in Example 7 of Section 6, if the linear recursion can be transformed into an iteration then we can evaluate f(v) using a *constant* number of processes only. Indeed, in this case, while the computation progresses we can reuse for the computation of new expressions, the processes which have been allocated to old expressions.

In order to understand the following Example 2, we need to introduce an irreflexive and transitive ordering > on the nodes of the symbolic graph of recursive calls which is assumed to have no loops: for any two nodes m and n, m > n holds iff the function call which labels m in the s-graph requires the computation of the function call which labels n.

**Example 2 (Towers of Hanoi)** Let us now consider the following familiar program for solving the Towers of Hanoi problem. Suppose that we have three pegs, say A, B, and C, and we want to move k disks from peg A to peg B, using C as an extra peg, by moving only one peg at a time. It should always be the case that smaller disks are placed over larger ones.

Let H be a function from Nat × $Peg^3$ to M* where M* is the monoid of moves freely generated by the set of possible moves M = {AB, BC, CA, BA, CB, AC}. The identity element of the monoid is 'skip' and ':' is the concatenation operation.

2.1 H(0,a,b,c) = skip
2.2 H(k+1,a,b,c) = H(k,a,c,b) : ab : H(k,c,b,a)

where the variables a, b, and c take *distinct* values in the set {A,B,C}, and for any two distinct values x and y in {A,B,C} the juxtaposition xy denotes a move in M.

We get the symbolic graph of recursive calls depicted in Fig. 5, where we have partitioned the nodes by *levels* according to the value of the first argument of H.

We now list the properties of that graph which suggest to us the definition of the auxiliary function to be introduced by tupling strategy [30] for obtaining a linear recursive program. Those properties are related to the function calls which are grouped together at level k–2, k–4, etc. (see the rectangles in Fig. 5).

Property i): we can express the function calls in the rectangle at level k+2 in terms of those at level k.

Property ii): there are three function calls in each rectangle.

Property iii): the initial function call H(k,a,b,c) can be expressed in terms of the functions in the rectangle at level k–2.

Each function triple in a rectangle is a *cut* of the s-graph, in the sense that if we remove the nodes of a cut together with their ingoing and outgoing edges, then we are left with two disconnected subgraphs, such that for each node m of the first subgraph and each node n of the second subgraph we have that: m > n.

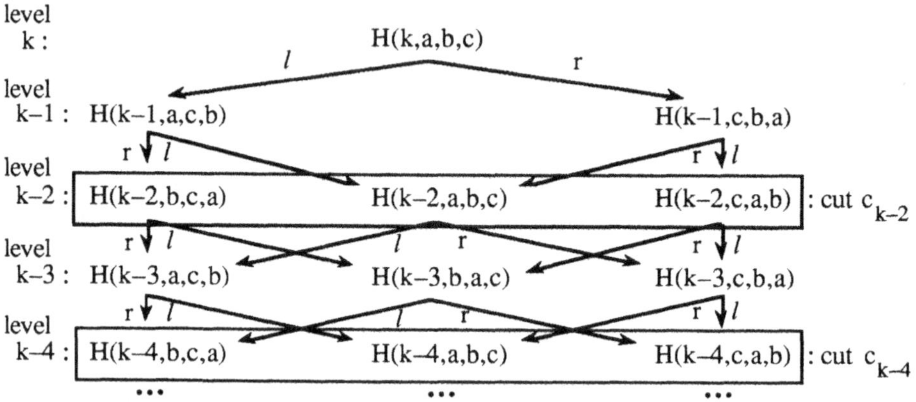


Figure 5: The symbolic graph of recursive calls of the function H(k,a,b,c).

We say that in the symbolic graph of recursive calls there exists a *progressive sequence of cuts* [30] iff there exists a sequence of cuts $\langle c_i \mid 0 \le i \rangle$ such that:

i) $\forall i > 0.\ c_{i-1}$ and $c_i$ have the same finite cardinality,
ii) $\forall i > 0.\ c_{i-1} \neq c_i$,
iii) $\forall i > 0.\ \forall n \in c_i.\ \exists m \in c_{i-1}.$ if $n \neq m$ then $m > n$, and
iv) $\forall i > 0.\ \forall m \in c_{i-1}.\ \exists n \in c_i.$ if $n \neq m$ then $m > n$.

From i) and ii) it follows that for all i such that i>0, neither $c_{i-1}$ is contained in $c_i$ nor $c_i$ in $c_{i-1}$. In intuitive terms, while moving along a progressive sequence of cuts from $c_{i-1}$ to $c_i$ we trade 'large' nodes for 'small' nodes.

Thus, given the s-graph of Fig. 5 where m > n is depicted by positioning the node m above the node n, we have, among others, the following cuts:

$c_{k-2}$ = {H(k–2,a,b,c), H(k–2,b,c,a), H(k–2,c,a,b)},

$c_{k-4}$ = {H(k–4,a,b,c), H(k–4,b,c,a), H(k–4,c,a,b)}, ..., and $\langle c_{k-2}, c_{k-4}, \ldots \rangle$ is a progressive sequence of cuts.

As explained in an earlier paper of ours [30], the existence of a progressive sequence of cuts, suggests the application of the *tupling strategy*. This means that we have to introduce a new function made out of the functions included in a cut. In our case we have:

$$t(k,a,b,c) =_{def} \langle\, H(k,a,b,c),\ H(k,b,c,a),\ H(k,c,a,b) \,\rangle.$$

The recursive equation for t(k,a,b,c) is:

2.3 t(k+2,a,b,c) = < H(k+2,a,b,c), H(k+2,b,c,a), H(k+2,c,a,b) > = {unfolding} =
= < H(k+1,a,c,b) : ab : H(k+1,c,b,a),
H(k+1,b,a,c) : bc : H(k+1,a,c,b),
H(k+1,c,b,a) : ca : H(k+1,b,a,c) > =
= < (u : ac : v) : ab : (w : cb : u), (v : ba : w) : bc : (u : ac : v),
(w : cb : u) : ca : (v : ba : w) > **where** <u,v,w> = t(k,a,b,c)
for k≥0.

Notice that when writing Equation 2.3 we may use the associativity of : and, for instance, we may write u : ac : v : ab : w : cb : u, instead of (u : ac : v) : ab : (w : cb : u). Thus, parallelism can be increased, so that we can compute (u : ac) in parallel with (v : ab) and (w : cb).

Then the value of H(k,a,b,c) can be expressed in terms of the tupled function t(k,a,b,c) as follows:

2.4 H(0,a,b,c) = skip
2.5 H(1,a,b,c) = = {unfolding} = skip : ab : skip = ab
2.6 H(k+2,a,b,c) = {unfolding} = H(k+1,a,c,b) : ab : H(k+1,c,b,a) = {unfolding} =
= H(k,a,b,c) : ac : H(k,b,c,a) : ab : H(k,c,a,b) : cb : H(k,a,b,c) =
= (u : ac : v) : ab : (w : cb : u) **where** <u,v,w> = t(k,a,b,c)
for $k \geq 0$.

In order to successfully apply the tupling strategy, we also need to preserve the termination properties of the initial program [22]. We will not present here a general theory for solving this problem. We will simply apply the following straightforward technique: we first look at the recursive definition of the function t, and we then search for suitable base cases which will make it to terminate as often as the given function H.

In our case, since t(k+2,a,b,c) is defined in terms of t(k,a,b,c), in order to ensure termination for the evaluation of t(k,a,b,c) for $k \geq 0$, we need to provide the equations for t(0,a,b,c) and t(1,a,b,c). We have:

2.7 t(0,a,b,c) = < skip, skip, skip >.
2.8 t(1,a,b,c) = < H(1,a,b,c), H(1,b,c,a), H(1,c,a,b) > = {unfolding} =
= < ab, bc, ca >.

Equations 2.4, 2.5, and 2.6 for the function H and Equations 2.3, 2.7, and 2.8 for the function t determine a linear recursive program which allows for a parallel execution by requiring a linear number of processes only, because: i) the number of components of the tupled function t is constant, and ii) the depth of the recursion linearly depends on the value of k. (Actually, as we will see in Example 7, we can compute the value of t(k,a,b,c) by using a constant number of processes only.) ■

With reference to Example 2 the evaluation of t(k+2,a,b,c) from the value of t(k,a,b,c) progresses as the parallel rewriting of the graph shown in Fig. 6 according to the rules R1, R2, R3, and MR. Synchronization among function calls take place *every second level*, in the sense that the three components of the function t must be tupled together at every second level (see Fig. 5).

The amount of parallelism during the evaluation of t(k+2,a,b,c) according to our model of computation, is limited by the existence of a unique copy of t(k,a,b,c), which is shared among the occurrences of the projection functions $\pi 1$, $\pi 2$, and $\pi 3$ (see Fig. 6). This fact may inhibit a fast evaluation of the function t. Thus, in order to increase parallelism we may make various copies of the value of t(k,a,b,c), once it has been computed.

Another solution which has been proposed in the literature [7] for the class of linear recursive programs derived from equation P, is as follows. One uses the initial non-linear program which requires an exponential number of processes, when the number of available processes is large, and when it tends to become small, one uses the transformed version which requires a linear number of processes only. The switching between the two modes of execution can be done at run time according to the actual needs of the computation.

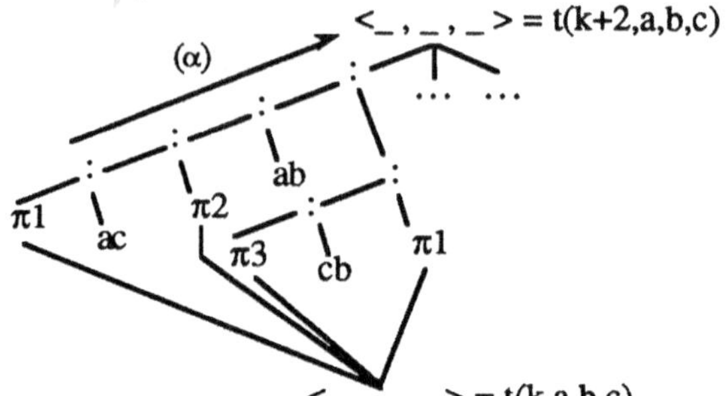


Figure 6: The evaluation of t(k+2,a,b,c) starting from t(k,a,b,c) according to Equation 2.3. Path (α) will be explained later.

# 4 Temporal and Spatial Synchronization

In order to improve the efficiency of the program we have derived in Example 2 we may want to avoid the redundancy which is present in Equation 2.3. Indeed, the three expressions u : ac : v, w : cb : u, and v : ba : w are computed twice.

Redundant computations of this form often occur in practice. We now show a method which reduces redundancy by using where-clauses and tuples of functions, and thus, enforcing some *synchronizations* among processes. Let us see how it works in our Towers of Hanoi Example, which we now revisit.

**Example 3 (Towers of Hanoi revisited: More Temporal Synchronizations)** By Equation 2.3 we synchronize the evaluation of the components of the function t at every second level. We may increase the so-called *temporal synchronization* by synchronizing the evaluation of the components of t *at every level*, and by doing so we may avoid the repeated evaluation of some identical subexpressions which is determined by Equation 2.3.

Indeed we have:

2.9 t(k+1,a,b,c) = < H(k,a,c,b) : ab : H(k,c,b,a), H(k,b,a,c) : bc : H(k,a,c,b), H(k,c,b,a) : ca : H(k,b,a,c) > =
= < p : ab : q, r : bc : p, q : ca : r > **where** <p,q,r> = t(k,a,c,b) for k≥0.

The graph which corresponds to the evaluation of t(k+1,a,b,c) starting from t(k,a,c,b) using Equation 2.9, is depicted in the lower part of Fig. 7.

The increase of temporal synchronization may eliminate some redundant computations at the expense of decreasing the amount of potential parallelism. This is indeed what happens in our case, because the processes needed for computing t(k+2,a,b,c) using Equation 2.3 are more than those needed when using Equation 2.9. In the graph of Fig. 6, in fact, we have more nodes than the ones in Fig. 7, and fewer nodes means that in general, fewer parallel rewritings may take place.

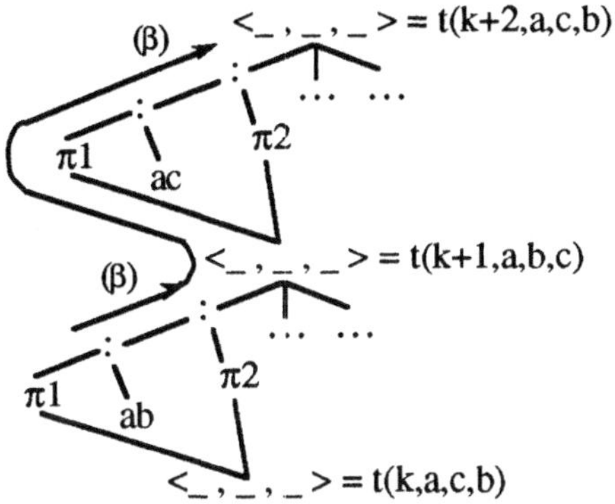


Figure 7: The evaluation of t(k+2,a,c,b) starting from t(k,a,c,b) according to Equation 2.9. Path (β) will be explained later.

Notice, however, that if we measure the computation time by the length of the longest sequence of concatenation operations ':' then the total amount of parallel time for computing t(k+2,a,b,c) does not change if we use Equation 2.3 or Equation 2.9. Indeed, both path (α) in Fig. 6 and path (β) in Fig. 7 have four concatenations. ■

There exists another technique for avoiding redundant computations. It consists in increasing the so-called *spatial synchronization* by increasing the number of function calls which are tupled together. This fact may create a dependency among the function calls occurring in a tuple, in the sense that in order to compute a component of a tuple, we need the value of another component of the same tuple. In that case we assume that if a component, say $t_i$, depends on another component, say $t_j$, of the same tuple then we evaluate $t_j$ before $t_i$, and this requirement may reduce the amount of parallelism. The following example will clarify the ideas.

**Example 4 (Towers of Hanoi revisited: More Spatial Synchronizations)**
We refer again to Example 2. For the evaluation of the function H(k,a,b,c) we may define the following function:

$$z(k+1,a,b,c) =_{def} < H(k+1,b,c,a),\ H(k+1,a,b,c),\ H(k,b,a,c),\ H(k,c,b,a) >,$$

with four components, instead of the function t(k,a,b,c) with three components. Thus, we have increased the spatial synchronization.

The function z which corresponds to the progressive sequence of cuts depicted in Fig. 8, is defined as follows:

2.10 $\quad z(1,a,b,c) = <\ bc,\ ab,\ skip,\ skip\ >$

2.11 $\quad z(2,a,b,c) = <\ ba : bc : ac,\ \ ac : ab : cb,\ \ ba,\ \ cb\ >$

2.12 $\quad z(k+3,a,b,c) = <\ H(k+3,b,c,a),\ H(k+3,a,b,c),\ H(k+2,b,a,c),\ H(k+2,c,b,a)\ > =$

$$\begin{aligned} &= <\ H(k+2,b,a,c) : bc : H(k+2,a,c,b),\\ &\qquad H(k+2,a,c,b) : ab : H(k+2,c,b,a),\\ &\qquad H(k+1,b,c,a) : ba : H(k+1,c,a,b),\\ &\qquad H(k+1,c,a,b) : cb : H(k+1,a,b,c)\ > = \{\text{unfolding}\} =\\ &= <\ (H(k+1,b,c,a) : ba : (H(k,c,b,a) : ca : H(k,b,a,c))) : bc :\\ &\qquad\qquad (H(k+1,a,b,c) : ac : H(k+1,b,c,a)), \end{aligned}$$

$$
\begin{aligned}
&(H(k+1,a,b,c) : ac : H(k+1,b,c,a)) : ab : \\
&\qquad ((H(k,c,b,a) : ca : H(k,b,a,c)) : cb : H(k+1,a,b,c)), \\
&H(k+1,b,c,a) : ba : (H(k,c,b,a) : ca : H(k,b,a,c)), \\
&(H(k,c,b,a) : ca : H(k,b,a,c)) : cb : H(k+1,a,b,c) > \quad \text{for } k\geq 0.
\end{aligned}
$$

Equations 2.10 and 2.11 are needed for preserving termination, because, as we will now see, z(k+3,...) depends on z(k+1,...) (see Equation 2.13 and Fig. 8 below).

Since H(k+3,b,c,a) depends on H(k+2,b,a,c) and H(k+3,a,b,c) depends on H(k+2,c,b,a), we first compute the values of H(k+2,b,a,c) and H(k+2,c,b,a) and we store them in the variables x and y, respectively. We get:

$$
\begin{aligned}
2.13\quad z(k+3,a,b,c) = (&< x : bc : (q : ac : p),\ (q : ac : p) : ab : y,\ x,\ y > \\
&\textbf{where } <x,y> = < p : ba : (s : ca : r),\ (s : ca : r) : cb : q >) \\
&\textbf{where } <p,q,r,s> = z(k+1,a,b,c) \qquad \text{for } k\geq 0.
\end{aligned}
$$

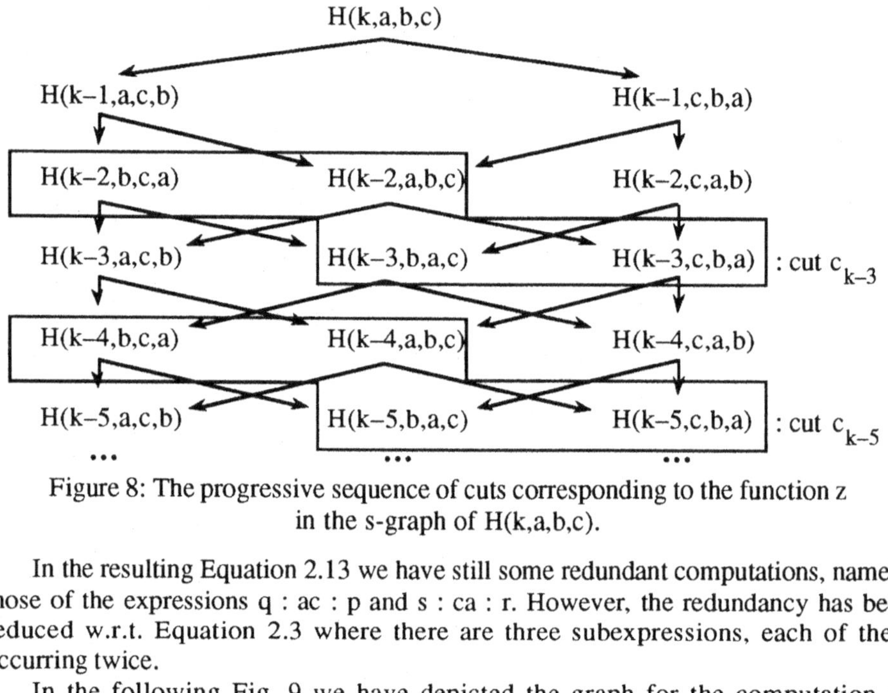


Figure 8: The progressive sequence of cuts corresponding to the function z in the s-graph of H(k,a,b,c).

In the resulting Equation 2.13 we have still some redundant computations, namely those of the expressions q : ac : p and s : ca : r. However, the redundancy has been reduced w.r.t. Equation 2.3 where there are three subexpressions, each of them occurring twice.

In the following Fig. 9 we have depicted the graph for the computation of z(k+3,a,b,c) according to Equation 2.13. The nodes (x) and (y) denote the subexpressions x and y of the where-clause of Equation 2.13.

The reduction of redundancy realized by Equation 2.13 w.r.t. Equation 2.3 has been obtained at the expense of increasing the *parallel computation time*, in the sense that the length of the longest sequence of concatenation operations ':' in Equation 2.13 is greater than the one in Equation 2.3. Indeed, we get from z(k+1,a,b,c) to z(k+3,a,b,c) through at most five concatenation operations (see path ($\gamma$) in Fig. 9), while we get from t(k,a,b,c) to t(k+2,a,b,c) through four concatenation operations only (see path ($\alpha$) in Fig. 6).

Now we show some facts about of the spatial synchronization which indicate that care is needed when choosing the functions to tuple together, because otherwise, program performances may not be improved. The Eureka Procedure for the application of the tupling strategy which we present in the following section, will indeed determine for us the good choices to be made.

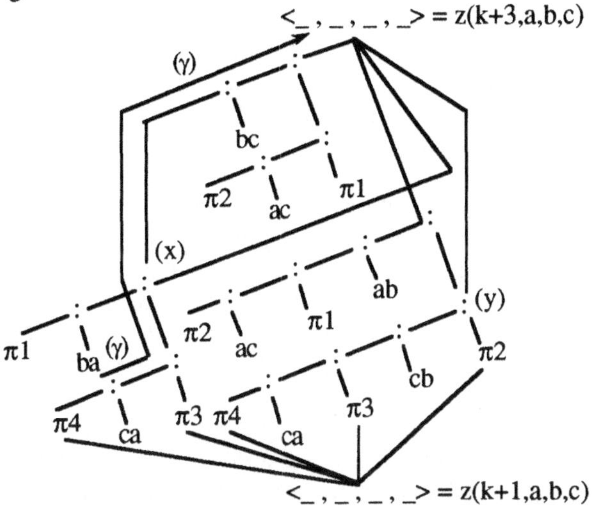


Figure 9: The evaluation of z(k+3,a,b,c) starting from z(k+1,a,b,c) according to Equation 2.13. Some arcs from π1 and π2 to z(k+1,a,b,c) have been omitted.

**Fact 6** Let us consider the program P which recursively defines the function f . i) The reduction of spatial synchronization below a certain threshold may determine an exponential number of repeated recursive calls of f (w.r.t. the depth of the recursion), while saving an exponential number of them. ii) The same amount of spatial synchronization may determine a linear or an exponential growth (w.r.t. the depth of the recursion) of the number of the recursive calls needed during the computation of the function f.

**Proof** Point i). Let us assume that we have defined the tupled function r(k+2) = <11,12> with two function calls only (see Fig. 10 (A)).

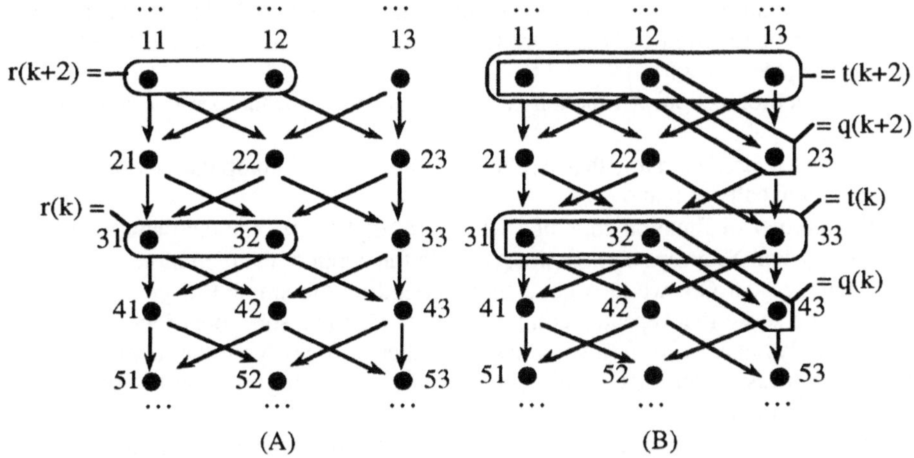


Figure 10: Two symbolic graphs of recursive calls similar to the one of H(k,a,b,c).

Since node 33 is a descendant of the node 13 in two different ways (see paths: 13, 22, 33 and 13, 23, 33), we have that at level k–2 node 53 is evaluated $2^2$ times. Analogously, we can show that node 73 is evaluated $2^3$ times, and so on.
Moreover, node 32 of the function r(k) is computed only once, while there are two paths leading to it from nodes of r(k+2). They are: 11, 21, 32 and 12, 23, 32. Thus, by using the tuple function r, we compute only once the value of the node 32, and an exponential saving of the recursive calls of f is achieved.

Point ii). Let us now consider Fig. 10 (B). We know already (see Example 2) that by tupling three function calls together and defining, for instance, the function t(k+2) = <11,12,13> we get a linear growth (w.r.t. the depth of the recursion) of the number of recursive calls of f. On the other hand, by tupling together three function calls (which do not constitute a cut) we may get an exponential growth. Indeed, during the computation of the tupled function q(k+2) = <11,12,23> recursively defined in terms of q(k), we compute twice the value of the node 33 (see the two paths: 11, 22, 33 and 23, 33). Thus, we will compute $2^2$ times the value of the node 53, and so on. ■

From Fact 6 it derives that when we tuple functions together for the optimal synchronization of parallel computations we need to determine both the *number* of function calls to be tupled and their *expressions*, otherwise we may fail to achieve the desired performances, that is, i) we may *not* get a linear growth (w.r.t. the depth of the recursion) of the number of processes while the computation progresses, and ii) we may *not* avoid redundancy, that is, we may cause some repeated computations of identical recursive calls.

# 5 Optimal Synchronization of Parallel Function Calls

In this section we will study the problem of finding an *optimally synchronized* parallel program which computes the least fixpoint function from X to Y defined by the program P: f(x) = **if** p(x) **then** a(x) **else** b(c(x), f(*l*(x)), f(r(x))), in the sense that:

G1) the minimal amount of spatial synchronization is required, that is, the minimal number of function calls of f are tupled together when we use the Eureka Procedure (see below),

G2) the synchronization of function calls does not increase the parallel computation time,

G3) there is no *redundancy*, that is, there are no repeated computations of identical recursive calls of f, and

G4) the number of function calls of f which are required for the computation of f(v) for any v in X, grows at most linearly with the depth of the recursion.

If we achieve goal G3, we also get a linear bound on the amount of *general redundancy*, in the sense that repeated computations of identical subexpressions (not recursive calls only) may occur at most a linear number of times w.r.t. the depth of the recursion. This result depends on the condition ii) concerning the construction of the computation graphs.

Since f is not linearly recursive, goal G4 cannot be achieved if during the computation of f(v) for some v in X, there are no repeated computations of identical calls of f. On the contrary, if the computation of f(v) does require multiple computations of identical calls of f and some suitable hypotheses are satisfied, then goals G1, G2, G3, and G4 can be achieved, as we will indicate below, by defining a new auxiliary function as a tuple of function calls of f.

Let us now develop our theory by introducing the following definitions which refer to a set of equations between words in $\Sigma^*$, where $\Sigma = \{l, r\}$. Those equations, as we already mentioned, may be used for representing identical function calls in an s-tree.

**Definition 7** Let E be a set of equations of the form s = t with $s, t \in \Sigma^*$. The *star closure* $E^*$ of the set E is the smallest set of equations including E, which is closed under the following rules:

i) (r: reflexivity) if $s \in \Sigma^*$ then $s = s \in E^*$,
ii) (s: symmetry) if $s = t \in E^*$ then $t = s \in E^*$, and
iii) (t: transitivity) if $s = t \in E^*$ and $t = u \in E^*$ then $s = u \in E^*$.

The *congruence closure* $E^c$, or *closure*, for short, of the set E is the smallest set of equations including E and closed under the rules i), ii), iii) above, and the following two rules:

iv.1) (*l*c: left congruence) if $a \in \Sigma$ and $s = t \in E^c$ then $a\,s = a\,t \in E^c$, and
iv.2) (rc: right congruence) if $a \in \Sigma$ and $s = t \in E^c$ then $s\,a = t\,a \in E^c$,

where juxtaposition denotes concatenation of words.

For any given set E of equations we define the *quotient set* $\Sigma^*/E^c$ as the set of $E^c$-equivalence classes of words in $\Sigma^*$ such that any two words, say u and v, are in the same class iff $u = v \in E^c$. ■

Notice that if we use the following rule:

iv) (c: congruence) if $s = t \in E^c$ and $u = v \in E^c$ then $s\,u = t\,v \in E^c$,

instead of rules iv.1 and iv.2, we get an equivalent definition of $E^c$. In particular, given any u, v, s, t in $\Sigma^*$ the equation s u = t v can be obtained from s = t, u = v, rule *l*c, and rule rc, as follows: s u = {some applications of rc from s = t} = t u = {some applications of *l*c} = t v.

We now assume that given the program P and an interpretation I of its basic operators, there exists a *finite* set E of equations between words in $\Sigma^*$, which is *characteristic* for <P,I> (or *characterizes* <P,I>), in the sense that the s-tree for <P,I> can be transformed into the corresponding s-graph by identifying any two nodes p and q of the s-tree with associated words u and v, respectively, iff u = v belongs to $E^c$.

We will not address here the problem of deciding whether or not for any given <P,I> there exists a finite characteristic set of equations and how it can be constructed. We only say that for many classes of programs such a finite set exists and it can be generated by performing some unfolding steps and proving some equalities. For instance, in the case of the Fibonacci function:

fib(x) = **if** x≤1 **then** 1 **else** fib(x–1) + fib(x–2)

by unfolding we get:

fib(x) = fib(x–1) + fib(x–2) =
= (fib((x–1)–1) + fib((x–1)–2)) + (fib((x–2)–1) + fib((x–2)–2)).

Now, if we show that for any natural number x we have that (x–1)–1 = x–2 then we derive the equation $l\!l$=r. It is not difficult to see that this equation allows us to derive from the s-tree of the function fib(x) the corresponding s-graph.

**Definition 8** Given the equation s = t, its *frontier* is denoted by F(s = t) and it is equal to max{L(s), L(t)}. Given a set E of equations we say that its *frontier* is max{F(s = t) | s = t ∈ E}. An equation s = t is said to be *balanced* iff L(s) = L(t). ■

**Definition 9** Given a set E of equations between words in $\Sigma^*$ and an integer k ≥ 0, the *set of nodes at level* k, denoted by V(k), is the set {[s] | [s] ∈ $\Sigma^*/E^c$ and k = min{L(t) | t ∈ [s]}} of $E^c$-equivalence classes of words. ■

Thus, an $E^c$-equivalence class [s] of words belongs to V(k) iff no word in [s] has length smaller than k and there exists in [s] a word of length k. Obviously, the set $\Sigma^*/E^c$ of equivalence classes of words is partitioned by the sets V(k) for k≥0. Thus, by Definition 5, we have that the set of nodes of any s-graph for the program P and an interpretation I of its basic operators, is partitioned by the sets V(k) for k≥0.

For simplicity, in what follows we will feel free to indicate the equivalence class of words without their square brackets. Thus, for instance, we will feel free to write: s ∈ V(k), instead of [s] ∈ V(k), and we will feel free to write 'word' instead of 'equivalence class of words' when no confusion arises.

**Definition 10** Given an s-graph G characterized by the set E of equations, the corresponding *reduced symbolic graph of recursive calls* (or *reduced s-graph*, for short) is the subgraph of G with the same set of nodes, that is, $\cup_{i\geq 0}$ V(i), and the following set of arcs: {<p,q> | ∀ node p, q. <p,q> is an arc of G and *if* p∈V(i) and q∈V(j) *then* i<j}. ■

As for s-graphs, also in the reduced s-graphs we identify each node with an element of $\Sigma^*/E^c$. Examples of reduced s-graphs are given in Fig. 13–17. Since a reduced s-graph has the same set of nodes of the corresponding s-graph, any set E of equations which characterizes a given s-graph also characterizes its reduced s-graph and viceversa.

Notice that if the equation $l$ = r is in the closure $E^c$ of the finite set E of equations characterizing a given reduced s-graph then the program P is linear recursive, and in a linear recursive program goals G1, G2, G3, and G4 are all achieved, as we will explain later (see Theorem 23).

We will assume that neither $l$ = ε nor r = ε is in $E^c$, because if one of these two equations is in $E^c$, we have: f(x) = $a_I$(x). Indeed, since we have assumed that for all x≠⊥ f(x) terminates, we have that for all x≠⊥, $p_I$(x) = true.

Therefore, we will assume that the equations in $E^c$ with frontier less than 2 are either {ε = ε, $l$ = $l$, r = r} or {ε = ε, $l$ = $l$, r = r, $l$ = r}.

We may now present some fundamental properties of the reduced s-graph for the program P, an interpretation I, and a finite characteristic set E of equations for <P,I>.

We need the following definitions and lemmas.

**Definition 11** Let C be an equivalence relation on $\Sigma^*$. Let us also assume that C is a congruence w.r.t. the concatenation operation. Given two sets A and B of C-congruence classes of words in $\Sigma^*$ and an element x in $\Sigma$, we write $A\ {}_x\!= B$ iff there exists a bijection, say $f_x$, from A to B such that for any [u] in A we have that $f_x([u]) = [x\,u]$. ■

This definition is well formed, because C is a congruence, and thus, if [u] = [v] then [xu] = [xv].

**Lemma 12** If we consider a set E of balanced equations between words in $\Sigma^*$ then for any $k \geq 0$, we have: $[s] \in V(k)$ iff $L(s) = k$.

**Proof** All equations in $E^c$ are balanced because all equations in E are balanced. Thus, a word of length k belongs to an $E^c$-equivalence class of words in V(k). ■

**Lemma 13** Let us consider a set E of equations between words in $\Sigma^*$. Let the frontier of E be $m \geq 1$. If u = v is in $E^c$ and $[u] \in V(L(u))$ and $L(u) > m$ then there exists a sequence $\langle w_1, w_2, \ldots, w_h\rangle$, with $h>1$, of words such that $w_1$ is u, $w_h$ is v, and for $i = 1,\ldots,h-1$ the equation $w_i = w_{i+1}$ is derived by one or more applications of left or right congruence rules to an equation of $E^*$.

**Proof** It is based on a normalization procedure of a proof of u = v. From the definition of V(L(u)) it follows that if $[u] \in V(L(u))$ then the length of any word w which is equal to u is not smaller than L(u). Thus, if there exists the sequence $\langle w_1, w_2,\ldots, w_h\rangle$ satisfying the hypothesis of this lemma, then for all i, $1 \leq i \leq h$, $L(w_i) \geq L(u)$.

Let us now consider a proof, say T, of the equation u = v starting from the equations in E. It can be represented as a tree (or term) built out of the symbols r, s, t, *lc*, and rc, denoting the application of the rules of reflexivity, symmetry, transitivity, left congruence, and right congruence, respectively. r has arity 0, because reflexivity has no premises, and t has arity 2, because transitivity has two premises. The other symbols have arity 1, because the corresponding rules have all one premise.

In the proof T the reflexivity rule can be applied only at the leaves. Without loss of generality we may assume that reflexivity is applied to derive the equation $\varepsilon = \varepsilon$ only ($\varepsilon$-reflexivity). All other equations of the form u = u for any $u \in \Sigma^+$, can be obtained from $\varepsilon = \varepsilon$ by applying the left congruence rule. Let U and V be subproofs of T. We can perform the following term (or proof) transformations:

1.1 $s(lc(U)) \Rightarrow lc(s(U))$, 1.2 $s(rc(U)) \Rightarrow rc(s(U))$,
2. $s(t(U,V)) \Rightarrow t(s(U),s(V))$, and
3.1 $lc(t(U,V)) \Rightarrow t(lc(U),lc(V))$, 3.2 $rc(t(U,V)) \Rightarrow t(rc(U),rc(V))$.

By doing so, we get a proof where all transitivity steps are performed after any other step, and the symmetry steps are performed before any left or right congruence step.

Now we can conclude that there exists a proof of each $w_i = w_{i+1}$ for $i = 1,\ldots,h-1$ whose last step is the application of a left or a right congruence because: i) for $i = 1,\ldots,h$, $L(w_i) \geq L(u) > m$, and ii) an equation in $E^c$ with frontier greater than m cannot be obtained by $\varepsilon$-reflexivity, symmetry, and transitivity from E. ■

**Definition 14** Given a set of equations between words in $\Sigma^*$ we say that 'Property ($\delta$) holds at level k' iff there exists $x \in \Sigma$ such that $V(k)\ {}_x\!= V(k+1)$. ■

We have the following theorem.

**Theorem 15** (*Balanced Equations*) Let us consider the program P, an interpretation I of its basic operators, and a finite characteristic set E of balanced equations for $\langle P,I\rangle$

between words in $\Sigma^+$. Let the frontier of E be m≥1. If Property (δ) holds at level p ≥ m–1 then Property (δ) holds at every level k, with k ≥ p.

**Proof** By hypothesis there exists $x \in \Sigma$ such that $V(p)\ _x\!= V(p+1)$. We will show that for any k ≥ p there exists a bijection $f_{x,k}$ from V(k) to V(k+1) such that for any [u] in V(k) we have: $f_{x,k}([u]) = [xu]$. The proof is by induction on k. The base case for k = p is obvious, because Property (δ) holds at level p. For the step case we will now show that: 1) $f_{x,k}$ is total, and $f_{x,k}$ is both 2) injective and 3) surjective.

*Point 1.* $f_{x,k}$ is total, because if s is in V(k+1) then x s is in V(k+2) by Lemma 12.

*Point 2.* We now show that $f_{x,k+1}$ is an injection from V(k+1) to V(k+2).

Assume $f_{x,k}$ is a bijection from V(k) to V(k+1) for k≥p. We have to show that for all $u, v \in V(k+1)$ if xu = xv then u = v.

Since xu = xv and L(xu) > m, by Lemma 13 there exists a sequence σ of words $<w_1, w_2, \ldots, w_h>$ for h>1, such that $w_1$ is xu, $w_h$ is xv, and for i = 1,…,h–1 the equation $w_i = w_{i+1}$ is derived by either a left congruence or a right congruence from an equation in $E^c$. We have that: for 1≤i≤h, $L(w_i) = k+2$, and $w_i$ belongs to the equivalence class xu of V(k+2). Let us consider the equation $w_i = w_{i+1}$ for each i = 1,…,h–1.

*Case 1.1.* If $w_i = w_{i+1}$ is derived by left congruence then $w_i = w_{i+1}$ is of the form: $as_i = as_{i+1}$ for some a in Σ and $s_i$, $s_{i+1}$ in $\Sigma^{k+1}$ with $s_i = s_{i+1}$. By Lemma 12 $s_i$ and $s_{i+1}$ belong to V(k+1).

*Case 1.2.* If $w_i = w_{i+1}$ is derived by right congruence then $w_i = w_{i+1}$ is of the form: $s_ib = s_{i+1}b$ for some b in Σ and $s_i$, $s_{i+1}$ in $\Sigma^{k+1}$ with $s_i = s_{i+1}$. By Lemma 12 $s_i$ and $s_{i+1}$ belong to V(k+1). Since $f_{x,k}$ is a bijection, by induction hypothesis we have that there exist $p, q \in \Sigma^k$ such that: $s_i = xp$, $s_{i+1} = xq$, and p = q. Hence, by right congruence we have: pb = qb. Thus, we have that the equation $w_i = w_{i+1}$ is of the form: xpb = xqb. By Lemma 12 both pb and qb belong to V(k+1).

In both cases 1.1 and 1.2, we have that $w_i = w_{i+1}$ is of the form: $at_i = at_{i+1}$ for some a in Σ and $t_i$, $t_{i+1}$ in $\Sigma^{k+1}$ with $t_i = t_{i+1}$. We also have that $t_i$ and $t_{i+1}$ belong to V(k+1). Then, by applying the transitivity property along the sequence σ, we get u = v.

*Point 3.* In order to show that $f_{x,k+1}$ is a surjection from V(k+1) to V(k+2) we have to show that: $\forall\, u \in V(k+2).\ \exists\, v \in V(k+1).\ u = xv$.

By Lemma 12 we have that: u = sz for some $s \in V(k+1)$ and z in Σ. Thus, u = xwz for some $w \in V(k)$ because $s \in V(k+1)$ and by induction hypothesis $f_{x,k}$ is a surjection from V(k) to V(k+1). The required v is wz. ■

Now we want to study the properties of the s-graphs for a different characteristic set of equations. For this purpose we generalize the Property (δ) of Definition 14 as follows.

**Definition 14*** Given a set E of equations between words in $\Sigma^*$ we say that 'Property (δ*) holds at level k' iff V(k) and V(k+1) can be partitioned into h (≥1) sets of $E^c$-equivalence classes of words such that: i) $V(k) = \bigcup_{1\le i\le h} V_i(k)$, ii) $V(k+1) = \bigcup_{1\le i\le h} V_i(k+1)$, and iii) for i = 1,…,h, there exists $x \in \Sigma$ such that $V_i(k)\ _x\!= V_i(k+1)$.■

We consider the graph Grid whose set of nodes is: $\{<x,y> \mid x \in \{l\}^*, y \in \{r\}^*\}$ and whose directed arcs are constructed as follows: for any node <x,y> there is an arc from <x,y> to <xl,y> and an arc from <x,y> to <x,yr> (see Fig. 11). As usual, we can introduce in Grid the relations of father, son, ancestor, and descendant nodes.

Given a word w in $\Sigma^*$ and an element $a \in \Sigma$, we denote by L(w,a) the number of a's occurring in w. Obviously, L(w) = L(w,$l$) + L(w,r). Let $a^n$ denote the word made out of n a's.

Each node <x,y> in Grid is identified with the *associated* equivalence class [xy] of words in the quotient set $\Sigma^*/\{lr = rl\}^c$. The corresponding equivalence relation is called *G-equivalence*. We have that: [xy] = { w | L(w,$l$) = L(x) and L(w,r) = L(y)} and thus, all words in [xy] have the same length, that is, L(x) + L(y). Conversely, for each [w] in $\Sigma^*/\{lr = rl\}^c$, [w] is identified with the node $<l^{L(w,l)},r^{L(w,r)}>$. Equality of nodes is the same as equality of the associated G-equivalence classes of words.

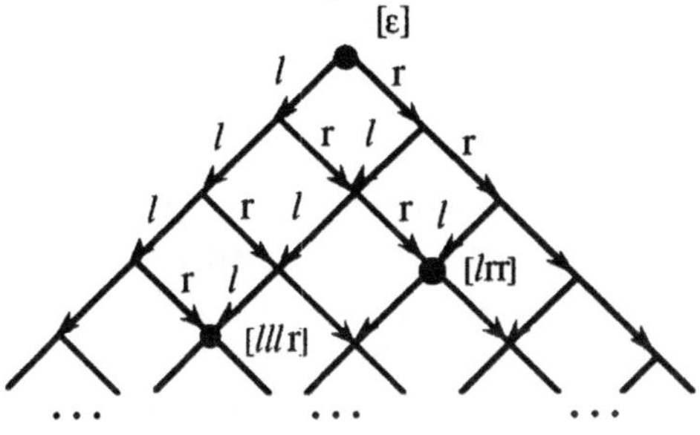


Figure 11: The graph Grid.
For h, k ≥ 0 the node $<l^h,r^k>$ is the G-equivalence class of words $[l^h r^k]$.

**Lemma 16** i) Given the set E = {$lr$ = r$l$} of equations, the corresponding V(k) for any k≥0, is the set of nodes [w] of Grid such that L(w) = k.
ii) Grid is the reduced s-graph characterized by the set E.
**Proof** It is by induction on k and it is based on the fact that u = v ∈ $\{lr = rl\}^c$ iff L(u,$l$) = L(v,$l$) and L(u,r) = L(v,r). ■

**Definition 17** The *dominated region* of a node [p] in Grid, denoted by Dom([p] ), is the subgraph of Grid whose set of nodes is {[pu] | u ∈ $\Sigma^*$}.

The *nodes to the left* (or *to the right*) of the node $[l^p r^q]$ for p, q ≥ 0, are those of the form $[l^i r^j]$, where i ≥ 0 and 0 ≤ j < q (or j ≥ 0 and 0 ≤ i < p) (see Fig. 12). ■

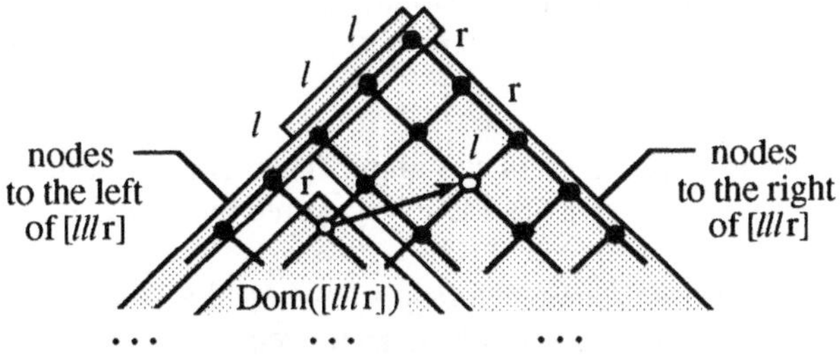


Figure 12: The translation from the node [$lllr$] to the node [$lrr$], and the nodes to the left and to the right of [$lllr$].

**Definition 18** Given any two nodes $[l^m r^n]$ and $[l^h r^k]$ in Grid, for m, n, h, k ≥ 0, the *translation* from the first node to the second one is the bijection from Dom($[l^m r^n]$) to Dom($[l^h r^k]$) which relates the node $[l^p r^q]$ to the node $[l^{p+h-m} r^{q+k-n}]$. ■

Notice that this definition is well formed. Indeed, if $[l^p r^q]$ is in Dom($[l^m r^n]$) then $m \leq p$ and $h \leq p+h-m$. Similarly, we have that $k \leq q+k-n$ and thus, $[l^{p+h-m} r^{q+k-n}]$ is in Dom($[l^h r^k]$).

In Fig. 12 we have depicted the translation from $[llr]$ to $[lrr]$ which relates the node $[l^p r^q]$ to $[l^{p-2} r^{q+1}]$ for any $p \geq 3$ and $q \geq 1$.

Given two graphs A and B, we denote by A – B the graph obtained by taking away from A the nodes and the arcs of B, and eliminating the arcs going to nodes of B.

Given two distinct nodes [u] and [v] of Grid, if we apply one or more times the translation from [v] to [u] with $L(v) \geq L(u)$, then each node in Dom([v]) is mapped into a node in Grid – Dom([v]).

**Theorem 19** Let E be the set $\{lr = rl,\ u = v\}$, where $1 \leq L(u) < L(v)$. The reduced s-graph characterized by E is the graph Grid – Dom([v]), where the nodes are considered to be $E^c$-equivalence classes, instead of G-equivalence classes.

**Proof** Since $L(u) \neq L(v)$ we have that $u = v \notin \{lr = rl\}^c$ and the G-equivalence classes of words [u] and [v] are associated with two distinct nodes in Grid. To prove the theorem it is enough to show that by considering all nodes as G-equivalence classes of words, for any given node [h] in Dom([v]) there exists a node [k] in Grid – Dom([v]) such that: 1) $h = k \in E^c$, and 2) $\min\{L(w) \mid w \in [h]\} > \min\{L(z) \mid z \in [k]\}$.

(This second part of the proof is needed for showing that when constructing the reduced s-graph characterized by E from Grid – Dom([v]) all arcs of Grid going into nodes of Dom([v]) can be deleted, because in any reduced s-graph there are no arcs from V(i) to V(j) if $i \geq j$.)

Point 1). Let us consider a node $[h] \in$ Dom([v]). We have that $[h] = [v l^p r^q]$ for some $p, q \geq 0$. By applying to [h] the translation from [v] to [u], we get the node $[u l^p r^q]$. This translation corresponds to the application of suitable left and right congruence steps to the equation u = v. Thus, $h = u l^p r^q \in E^c$. If the node $[u l^p r^q] \notin$ Dom([v]) then it is the node [k] of Grid – Dom([v]) we want to find. Otherwise, $[u l^p r^q] \in$ Dom([v]) and thus, $[u l^p r^q] = [v l^i r^j]$ for some $i, j \geq 0$. Then, by applying one or more times the translation from [v] to [u], we eventually get a node in Grid – Dom([v]) which is equal to [h].

Point 2). Let us consider a node $[h] \in$ Dom([v]). Since $L(u) < L(v)$, the translation from [v] to [u] relates any word w in the G-equivalence class associated with the node [h] to a word of length less than L(w). ■

Notice that in the hypotheses of the above Theorem 19 we have assumed that $L(u) < L(v)$, because the case of $L(u) = L(v)$ is covered by Theorem 15.

**Corollary 20** Let E be the set $\{lr = rl,\ u = v\}$, where $1 \leq L(u) < L(v)$. The set V(k) of nodes at level k in the reduced s-graph characterized by E, for any $k \geq 0$, is the set of nodes which are obtained by: i) identifying each node of Grid – Dom([v]) with the $E^c$-equivalence class of words, and ii) considering only those $E^c$-equivalence classes in Grid – Dom([v]) whose shortest representative has length k. ■

**Theorem 21** (*Commutativity*) Let us consider the program P, an interpretation I of its basic operators, and the characteristic set $E = \{lr = rl,\ u = v\}$ of equations for <P,I>

such that: i) $1 \leq L(u) < L(v)$, and ii) the node [u] is not an ancestor of [v] in Grid, where [x] denotes the G-equivalence class of words.
Then there exists $h \in \{1,2\}$ such that for all $k \geq m-1$ Property ($\delta$*) holds at level k in the reduced s-graph characterized by E.

**Proof** We will provide an informal proof by referring to Fig. 13 below. The bijection $g_k$ from V(k) to V(k+1) for any $k \geq m-1$ is defined as follows: it maps the nodes to the left of [v] at level k (that is, the nodes {1,2,3} for k=m) to their left-sons at level k+1 (that is, the nodes {7,8,9}), and it maps the nodes to the right of [v] at level k (that is, the nodes {4,5,6}) to their right-sons at level k+1 (that is, the nodes {10,11,12}). It is easy to see that for all $k \geq m-1$, V(k) can be considered as the union of the two disjoint sets $V_1(k)$ and $V_2(k)$, and we have that $g_k$ is made out of two bijections: the first one from $V_1(k)$ to $V_1(k+1)$ which maps any node [b] to [b*l*], and the second one from $V_2(k)$ to $V_2(k+1)$ which maps any node [b] to [br]. Since [b*l*]=[*l*b] and [br]=[rb], we have that for every $k \geq m-1$, $V_1(k)\ {}_l{=}\ V_1(k+1)$ and $V_2(k)\ {}_r{=}\ V_2(k+1)$.
The cases when there are no nodes to the left or to the right of [v] are analogous, and in those cases the function $g_k$ is made out of one bijection which maps nodes either to their left-sons or to their right-sons only. Thus, Property ($\delta$*) holds for h = 1 or 2. ■

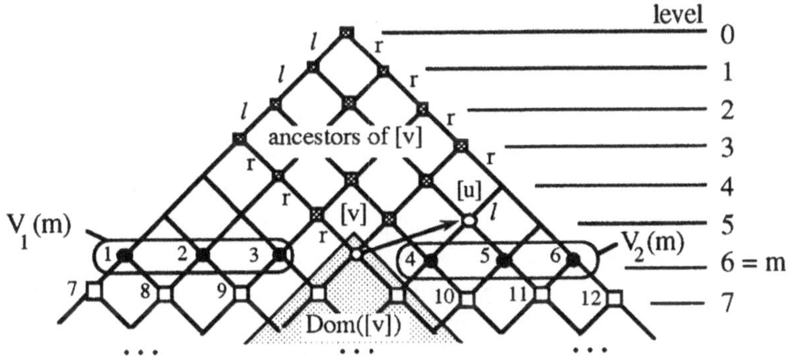


Figure 13: The reduced s-graph characterized by {*l*r = r*l*, u = v}, where u is *l*rrr and v is *lll*rr. The nodes in Dom([v]) do not exist.

The following procedure can be used to derive an optimally synchronized parallel program for evaluating the function f(x). The proof of correctness and optimality of this procedure is based on the above Theorems 15 and 21, and Theorems 22 and 23 which will be given below.

**EUREKA Procedure**

**Input** 1) The program P: f(x) = **if** p(x) **then** a(x) **else** b(c(x), f(*l*(x)), f(r(x))), with an interpretation I of its basic operators, defining the least fixpoint function $f_{fix} : X \rightarrow Y$, also denoted by f, when no confusion arises.

2) A finite set E of equations for <P,I> between words in $\{l,r\}^*$ which characterizes the reduced s-graph of f(x). Let the frontier of E be the integer $m \geq 1$. (*Linear Case*) $l = r \in E^c$. (*Balanced Case*) $l = r \notin E^c$ and each equation in E is balanced. (*Commutative Case*) E = {*l*r = r*l*, u = v}, where: i) $1 \leq L(u) < L(v)$, and ii) there is no z in $\{l,r\}^+$ such that u z is v.

**Output** an optimally synchronized parallel program for computing f(x).

**Method** If $l = r \in E^c$ (*Linear Case*) then an optimally synchronized parallel program for f is: f(x) = **if** p(x) **then** a(x) **else** b(c(x),z,z) **where** z = f($l$(x)). Otherwise we perform the following steps.

1.a (*Balanced Case*) We compute the sets V(i), for i≥0, and we look for a level k ≥ m−1, such that Property (δ) holds at that level. We introduce the auxiliary function $t(x) =_{def} < f(u_{1I}(x)),\ldots, f(u_{qI}(x)) >$ where $\{[u_1],\ldots, [u_q]\} = V(k)$. We then obtain the recursive definition of t(x) by:
   i) unfolding once each call of f in t(x) such that we obtain the following calls of f only: $f(v_{1I}(x)), \ldots, f(v_{qI}(x))$,
   ii) using a where-clause whose bound variables $<z_1,\ldots,z_q>$ are equal to $< f(v_{1I}(x)), \ldots, f(v_{qI}(x)) >$ with $\{[v_1],\ldots, [v_q]\} = V(k+1)$, and
   iii) replacing $< f(v_{1I}(x)),\ldots, f(v_{qI}(x)) >$ by $t(d_I(x))$ where d is the element of $\{l,r\}$ such that $V(k)\ {}_d{=}\ V(k+1)$. (This step is called *folding*.)

1.b (*Commutative Case*) We compute the sets V(i), for i = 0,…,m. Let us assume, for simplicity reasons, that Property (δ*) holds at level k=m−1 for h=2, that is, $V(k) = V_1(k) \cup V_2(k)$, $V(k+1) = V_1(k+1) \cup V_2(k+1)$, $V_1(k)\ {}_{d1}{=}\ V_1(k+1)$, and $V_2(k)\ {}_{d2}{=}\ V_2(k+1)$. We introduce the auxiliary function $t(x_1,x_2) =_{def} < f(u_{1I}(x_1)), \ldots, f(u_{jI}(x_1)), f(u_{j+1I}(x_2)), \ldots, f(u_{qI}(x_2)) >$ where $\{[u_1],\ldots, [u_j]\} = V_1(k)$ and $\{[u_{j+1}],\ldots, [u_q]\} = V_2(k)$. We then obtain the recursive definition of $t(x_1,x_2)$ by:
   i) unfolding one or more times each call of f in $t(x_1,x_2)$ such that we obtain the following calls of f only: $f(v_{1I}(x_1)),\ldots, f(v_{jI}(x_1)), f(v_{j+1I}(x_2)), \ldots, f(v_{qI}(x_2))$,
   ii) using a where-clause whose bound variables $<z_1,\ldots,z_j, z_{j+1},\ldots,z_q>$ are equal to $< f(v_{1I}(x_1)),\ldots, f(v_{jI}(x_1)), f(v_{j+1I}(x_2)), \ldots, f(v_{qI}(x_2)) >$ with $\{[v_1],\ldots, [v_j]\} = V_1(k+1)$ and $\{[v_{j+1}],\ldots, [v_q]\} = V_2(k+1)$, and
   iii) replacing $< f(v_{1I}(x_1)),\ldots, f(v_{jI}(x_1)), f(v_{j+1I}(x_2)),\ldots, f(v_{qI}(x_2)) >$ by $t(d1(x_1),d2(x_2))$. (This step is called *folding*.)

   If h = 1 the above steps i), ii), and iii) are like in the Balanced Case and the tuple function t to be introduced has one argument only.

2. Then by performing some unfolding steps, we express f(x) in terms of the function calls which are the q components of the tuple function t which corresponds to the elements of V(k).
3. Finally, we add to the linear recursive definition of the function t and the expression of f(x) in terms of t, suitable base cases both for the expression of f(x) and the definition of t, so that for any v in X the termination of the evaluation of f(v) is preserved. These base cases can be derived by performing some unfolding steps. ■

**Remarks** 1) Given the set E of equations, in order to check whether or not $l = r$ belongs to $E^c$ we can use the efficient algorithm given in [14].

2) For the computation of the sets V(i), for i ≥0, we recall that: $V(0) = \{[\varepsilon]\}$, $V(1) = \{[l], [r]\}$, and we can compute V(i+1) from V(0), …,V(i) as follows: a) take $V_{i+1} = \varnothing$, b) for each word x of length i+1 check whether or not it belongs to an $E^c$-equivalence class of the set $V(0) \cup \ldots \cup V(i) \cup V_{i+1}$ (to this purpose we can use again the algorithm given in [14]), and in the affermative case do nothing, otherwise add [x] to $V_{i+1}$. V(i+1) is the final value of $V_{i+1}$.

3) In the Balanced Case the Eureka Procedure may not terminate.

4) The Eureka Procedure extended in the obvious way, can also be used for produc-

ing an optimally synchronized parallel program when the arity of f is greater than 1. ■

The name 'Eureka' to the above procedure comes from the fact that t(x) is the function to be introduced during the so-called *eureka steps*, according to the terminology of [Burstall-Darlington 77]. In the following section we will give some examples of application of this procedure.

**Theorem 22** (*Success of the Unfolding Steps*) In the Eureka Procedure we can perform the unfoldings mentioned at steps 1.a.i), 1.b.i), and 2.
**Proof** Step 1.a.i (Balanced Case). We have to show that by unfolding once each component of the tuple $< f(u_{1I}(x)),\ldots, f(u_{qI}(x)) >$ which defines t(x), we get an expression in terms of the function calls $f(v_{1I}(x)),\ldots, f(v_{qI}(x))$ only. Indeed, the left and right sons of the function calls which label the nodes of the reduced s-graph at level k are all at level k+1 (by Lemma 12).
Step 1.b.i (Commutative Case). We have to show that by some unfolding steps we can express each component of the tuple $< f(u_{1I}(x_1)),\ldots, f(u_{jI}(x_1)), f(u_{j+1I}(x_2)), \ldots, f(u_{qI}(x_2)) >$ which defines $t(x_1,x_2)$, in terms of $f(v_{1I}(x_1)), \ldots, f(v_{jI}(x_1)), f(v_{j+1I}(x_2)), \ldots, f(v_{qI}(x_2))$ only. Thus, by referring to the reduced s-graph and saying 'node p' instead of 'the function call $f(p_I(x))$', we have to show that by performing some unfolding steps, each node in V(m–1) can be expressed in terms of the nodes in V(m). The only problem arises from the fact that by unfolding once the nodes in V(m–1) we get the call $f(v_I(x))$ corresponding to the node [v], which does not belong to V(m) because u = v and L(u) < m. However, since [v] = [u] and [u] is not an ancestor of [v], by unfolding the function call $f(v_I(x))$, that is, node [u], we get an expression in terms of the nodes in V(m) (see for instance, Fig. 13 where by unfolding [u] we get the nodes 4 and 5 in V(m) with m = 6).
Step 2. We have to show that by performing some unfolding steps, the topmost node of the reduced s-graph can be expressed in terms of the nodes in V(k). Let us first notice that in the Commutative Case and in the Balanced Case all equations holding between words of length at most k are balanced. (This is obvious in the Balanced Case, while in the Commutative Case only the balanced equation $lr = rl$ is applicable, because k = m–1.) Thus, by Lemma 12, if we unfold once each call in V(i) with $0 \le i < k$, we get the calls in V(i+1). Hence, by performing some unfolding steps, the topmost node of the reduced s-graph can be expressed in terms of the nodes in V(k).■

The following theorem shows that the Eureka Procedure produces optimally synchronized parallel programs.

**Theorem 23** (*Optimality of the Eureka Procedure*) Given the program P and an interpretation I, if $l = r \in E^c$ (Linear Case) where E is a characteristic set of equations for <P,I>, then an optimally synchronized parallel program is: f(x) = **if** p(x) **then** a(x) **else** b(c(x),y,y) **where** y = f(*l*(x)).
ii) In the Balanced Case and the Commutative Case, if by performing some unfolding steps we can express f(x) in terms of $z(x) =_{def} < f(w_{1I}(x)), \ldots, f(w_{bI}(x)) >$ where $\{[w_1],\ldots, [w_b]\} \subseteq V(j)$, and z(x) in terms of $z(a_I(x)) = < f(w_{1I}a_I(x)), \ldots, f(w_{bI}a_I(x))>$ where $a \in \Sigma^s$, $s > 0$, and $\{[aw_1],\ldots, [aw_b]\} \subseteq V(j+s)$, then $|V(k)| \le |V(j)|$, where V(k) is the level considered in the Eureka Procedure.
**Proof** i) The reduced s-graph characterized by the set E of equations such that $l = r \in E^c$,

is a sequence of the nodes $\{n_0, n_1, \ldots\}$ with the arcs $\{<n_0,n_1>, <n_1,n_2>, \ldots\}$. Such a sequence is finite iff in E there exists an equation which is not balanced. It is easy to see that goals G1, G2, G3, and G4 are all achieved by using the linear recursive program of the form: f(x) = **if** p(x) **then** a(x) **else** b(c(x),y,y) **where** y = f(*l*(x)). In particular, the spatial synchronization is the minimal one, because one function call only is synchronized.
ii) Let us first notice that for computing the topmost node of a reduced s-graph we need all nodes of any given level V(i) for $i \geq 0$. Thus, by our hypotheses on the function z we have that: $\{[w_1],\ldots,[w_b]\} = V(j)$, $|V(j)| = b$, $\{[aw_1],\ldots,[aw_b]\} = V(j+s)$, and $|V(j+s)| \leq b$, because it may be the case that $w_c \neq w_d$ and $aw_c = aw_d$ for some c, d in $\{1,\ldots,b\}$. We also have that in general $\{[a^n w_1],\ldots,[a^n w_b]\} = V(j+ns)$ and $|V(j+ns)| \leq b$ for each $n \geq 0$.
By Theorems 15 and 21 for all $k_1 \geq k$, $|V(k)| = |V(k_1)|$. Thus, if we take $j+ns \geq k$ we get: $|V(k)| = |V(j+ns)| \leq b = |V(j)|$. ■

In the above theorem we have restricted ourselves to the case where z has one argument only. A similar result can be established also in the case where z has more than one argument.

From Theorems 22 and 23 we may conclude that in the Balanced and Commutative cases goal G1 is achieved. Goal G2 is achieved, because the recursion of the tuple function introduced by the Eureka Procedure, is not deeper than the one of the deepest recursive call of f. Goal G3 is achieved because by construction, in the reduced s-graph there is only one node for each distinct recursive call of f. Also goal G4 is achieved because the definition of the tuple function which is determined by the Eureka Procedure, is linear recursive.

The optimal parallel program derived by the Eureka Procedure can often be further improved as shown by Example 7 in the following section. Indeed, in many cases it is possible to transform a linear recursive program into an iterative one which has the same time performances and uses a constant number of memory cells only (and thus, a constant number of processes, assuming that we need one process for the parallel updating of one memory cell).

# 6 Examples of Synthesis of Optimal Parallel Programs

In this section we will present some examples of application of the Eureka Procedure for the derivation of optimally synchronized parallel programs which compute various classes of functions defined by the program P. In what follows we will also omit, for simplicity, the explicit reference to the interpretation function I, and for instance, we will write p(x), instead of $p_I(x)$.

**Example 5 (Commutative Case: Common Generator Redundant Equations)** [9].
Let us suppose that in the program P there exists a function v(x), called *common generator*, and two positive integers h and k, such that:

i) $h \le k$, and ii) $l(x)=v^h(x)$ and $r(x)=v^k(x)$ for all x in X, where $v^0(x)$ denotes x and $v^{n+1}(x)$ for any $n \ge 0$ denotes $v(v^n(x))$. In particular, this implies that $l(r(x))=r(l(x))$ for all x in X.

Let us assume that D is the least common multiple (l.c.m.) of h and k. This implies that there exist two positive integers p and q such that $p \ge q > 0$ and $D = p \times h = q \times k$.

If h=k then the given equation for f(x) is linear recursive and it is already optimally synchronized, and thus, in what follows we will assume that $h < k$ and hence, $p > q$.

A set E of equations which characterizes the reduced s-graph of f(x) is: $\{lr = rl,\ l^p = r^q\}$. It has frontier p (>1). Notice also that $[l^p]$ is not an ancestor of $[r^q]$, because q>0. By means of the Eureka Procedure (Commutative Case) we obtain the function with p components (see Fig. 14):

$$t(x) =_{def} < f(v^{(p-1)h}(x)),\ f(v^{(p-2)h+k}(x)),\ f(v^{(p-3)h+2k}(x)),\ \ldots,\ f(v^{h+(p-2)k}(x)),\ f(v^{(p-1)k}(x)) >.$$

As ensured by Theorem 21, it is easy to check that in our case Property ($\delta^*$) holds at level p–1 for h = 1 (see levels p–1 and p of Fig. 14) and $V(p-1) \,_r\!= V(p)$.

In Fig. 14 we have used the number z to indicate the node whose label is the function call $f(v^z(x))$. In Fig. 14 and in the following ones, some nodes, that is, their associated $E^c$-equivalence classes of words, have been decorated with crosses. Those nodes and their ingoing arcs *do not exist* in the reduced s-graphs, and they have been depicted simply to indicate that they are identified with other nodes which occur nearer to the top node. In particular, in Fig. 14 the node ph occurs in the sequence ($\alpha$) = <k, 2k, …, (p–1)k> of nodes, because: ph = qk, k>h>0, and p>q>0. Analogously, the node ph+k occurs in <2k, …, pk>.

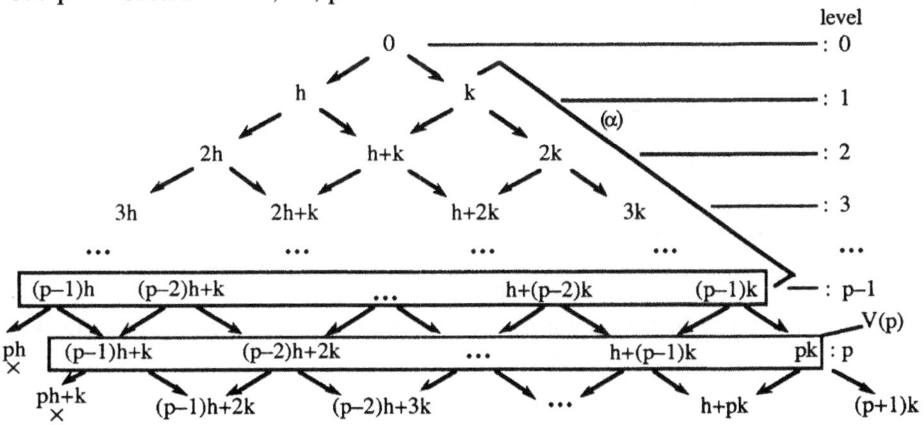


Figure 14: Reduced s-graph of f(x) in the case of common generator redundancy from level 0 to level p+1. The number z stands for the function call $f(v^z(x))$. Crossed nodes and their ingoing arcs do not exist.

Functions defining *linear recurrence relations* belong to the class of functions for which there exists a common generator. Let us consider the following example.

5.1 $d(0) = 1,\quad d(1) = 2,\quad d(2) = 0,$
5.2 $d(n+3) = d(n+2)+2d(n)\quad$ for $n \ge 0$.

In this case we have: $l = \lambda n.n-1$, $r = \lambda n.n-3$, $v = \lambda n.n-1$, $h = 1$, $k = 3$, and $D = \text{l.c.m.}\{1,3\} = 3$. A set E of equations which characterizes the reduced s-graph is: $\{lr = rl,\ lll = r\}$. (Notice that $lr = rl$ is implied by $lll = r$). The frontier of E is $m = 3$.

By applying the Eureka Procedure (Commutative Case) we have:
$V(0) = \{[\varepsilon]\}$, $V(1) = \{[l], [r]\}$, $V(2) = \{[ll], [lr], [rr]\}$, and
$V(3) = \{[llr], [lrr], [rrr]\}$. The function to be introduced is (see level 2 of Fig. 15):

$$t(n) =_{def} < d(n-2), d(n-4), d(n-6) >.$$

By expressing the components of the function t in terms of the components one level below (see Fig. 15), that is, t(n) in terms of t(n–3), we get:

5.3 $t(n) = < d(n-2), d(n-4), d(n-6) > = \{\text{unfolding}\} =$
$= < 3\,u + 4\,v + 4\,w,\ u + 2\,v,\ v + 2\,w >$ **where** $<u,v,w> = t(n-3)$ for $n \geq 9$.

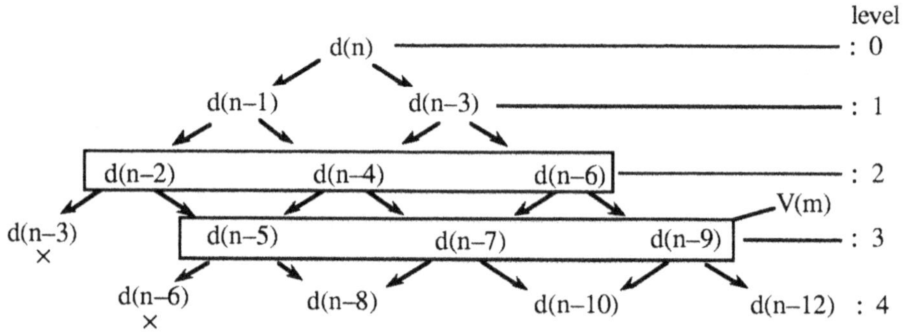


Figure 15: Reduced s-graph from level 0 to level 4 for the function d(n).

The constraint '$n \geq 9$' comes from the fact that the third component of t(n–3) is d(n–9) and the argument of d should not be negative. Then, since t(n) is defined in terms of t(n–3), in order to ensure termination we need to define the following three consecutive values of the function t:

5.4 $t(6) = < d(4), d(2), d(0) > = <6, 0, 1>$
5.5 $t(7) = < d(5), d(3), d(1) > = <6, 2, 2>$
5.6 $t(8) = < d(6), d(4), d(2) > = <10, 6, 0>$.

We now express the function d(n) (see level 0 of Fig. 15) in terms of the function t(n) (see level 2 of Fig. 15). We get:

5.7 $d(n) = d(n-1)+2d(n-3) = d(n-2) + 4\,d(n-4) + 4\,d(n-6) =$
$= u + 4\,v + 4\,w$ **where** $<u,v,w> = t(n)$ for $n \geq 6$.

Since d(n) is defined in terms of t(n) and t(n) is defined for $n \geq 6$, in order to ensure termination we have to provide the values of the function d for $n = 0,\ldots,5$:

5.8 $d(0) = 1$, $d(1) = 2$, $d(2) = 0$,
5.9 $d(3) = d(2)+2d(0) = 2$, $d(4) = d(3)+2d(1) = 6$, $d(5) = d(4)+2d(2) = 6$.

The final program made out of Equations 5.3 through 5.9 is linear recursive and it computes the same function defined by the given Equations 5.1 and 5.2.

As ensured by the results presented in the previous section, the synchronization of three function calls in the tuple function t is an optimal one. Indeed, if we tuple together only two function calls, we cannot compute the value of d(n) at level 0 without performing redundant evaluations of the function d (recall the proof of Theorem 23). We also have that the parallel evaluation of the three components of t(n) does not require repeated evaluations of identical function calls, it does not increase the parallel computation time, and it requires a linear number of computing processes only. ■

**Example 6 (Commutative Case: the Impatient Commuter Function)** [9]. Let us consider the definition of the function f(x) as in Example 5 and let us assume that for any x we have: i) $l(r(x)) = r(l(x))$ and ii) $l^p(x) = r^q(x)$ for some $p > q > 0$.

A set E of equations which characterizes the reduced s-graph of f(x) is (as in Example 5): $\{lr = rl,\ l^p = r^q\}$. It has frontier p (>1). Notice also that $[r^q]$ is not an ancestor of $[l^p]$, because q>0. Thus, we have that also the reduced s-graph of f(x) is like the one of Example 5.

This class of programs properly contains the class described in Example 5, because it is not required the existence of the common generator function. In particular, the following function f(i,h) satisfies the hypotheses i) and ii) above, but it is not an instance of the class of Example 5 [9]:

6.1 f(i,h) = **if** i≥k **then** a(i) **else** b(i, h, f(i+1,h), f(i+2,g(h))),

where ∀ x. g(g(x)) = x. In this case we have:
$l$(i,h) = <i+1,h> and r(i,h) = <i+2,g(h)>. Thus, we have: $llll$(i,h) = rr(i,h).

A set E of equations which characterizes the reduced s-graph of f(i,h) is: $\{lr = rl,\ llll = rr\}$. It has frontier m = 4. [rr] is not an ancestor of [$llll$]. By applying the Eureka Procedure (Commutative Case) we have:

V(0) = { [ε] }, V(1) = { [$l$], [r] },
V(2) = { [$ll$], [$l$r], [rr] }, V(3) = { [$lll$], [$ll$r], [$l$rr], [rrr] }, and
V(4) = { [$lll$r], [$ll$rr], [$l$rrr], [rrrr] }.

It is easy to see that Property (δ*) holds at level m−1 (see levels 3 and 4 of Fig. 16) for h = 1. In this case the function to be introduced is (see level 4 of Fig. 16):

$$t(i,h) =_{def} < f(i+3,h), f(i+4,g(h)), f(i+5,h), f(i+6,g(h)) >.$$

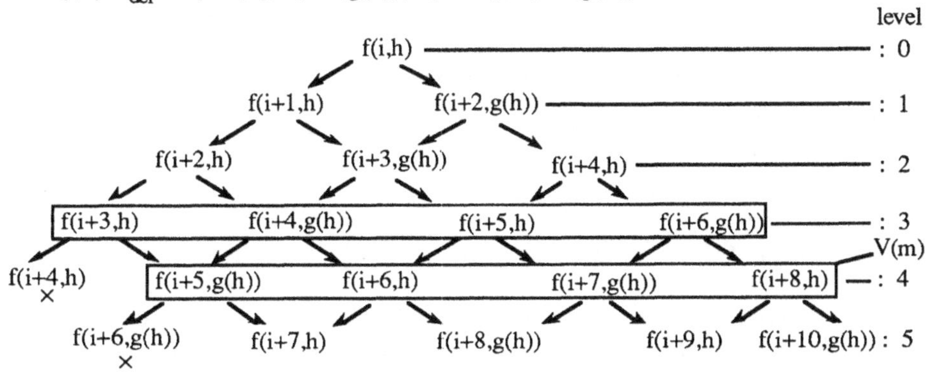


Figure 16: Reduced s-graph from level 0 to level 5 for the function:
f(i,h) = **if** i≥k **then** a(i) **else** b(i, h, f(i+1,h), f(i+2,g(h))).

Let us introduce the following abbreviations:

B(i, y, z) for **if** i≥k **then** a(i) **else** b(i, h, y, z) and
$\underline{B}$(i, y, z) for **if** i≥k **then** a(i) **else** b(i, g(h), y, z).

By expressing the components of t(i,h) at a given level in terms of the components at the level below (see Fig. 16), that is, t(i,h) in terms of t(i+2,g(h)), we get:

6.2 t(i,h) = < B( i+3, B( i+4, B(i+5, v, w), $\underline{B}$(i+6, w, z)), u),
$\underline{B}$(i+4, u, v), B(i+5, v, w),
$\underline{B}$(i+6, w, z) > **where** <u,v,w,z> = t(i+2,g(h)).

We now express the function f(i,h) (see level 0 of Fig. 16) in terms of the function t(i,h) (see level 3 of Fig. 16). We get:

6.3 f(i,h) = B( i, B( i+1, B(i+2, u, v), $\underline{B}$(i+3, v, w)),
$\underline{B}$( i+2, $\underline{B}$(i+3, v, w), B(i+4, w, z))) **where** <u,v,w,z> = t(i,h).

The final program, that is, Equations 6.2 and 6.3, is linear recursive and it computes the function f(i,h) defined by Equation 6.1. Equations 6.2 and 6.3 achieve the optimality goals indicated at the beginning of the previous section. ■

**Example 7 (Balanced Case: Towers of Hanoi)** An optimally synchronized parallel program is derived as follows. Let us consider again Equations 2.1 and 2.2 of Example 2. We have:

7.1 H(k,a,b,c) = **if** k=0 **then** skip **else** H(k–1,a,c,b) : ab : H(k–1,c,b,a)

We also have: *l*(k,a,b,c) = <k–1,a,c,b> and r(k,a,b,c) = <k–1,c,b,a>,
A set E of balanced equations characterizing the reduced s-graph (see Fig. 17) is: { *ll* = rr, *lrl* = r*lr* }. The frontier of E is m = 3.

By applying the Eureka Procedure (Balanced Case) we have:
V(0) = { [ε] }, V(1) = { [*l*], [r] }, V(2) = { [*ll*], [*l*r], [r*l*] }, and
V(3) = { [*l*rr], [*lrl*], [r*ll*] }.

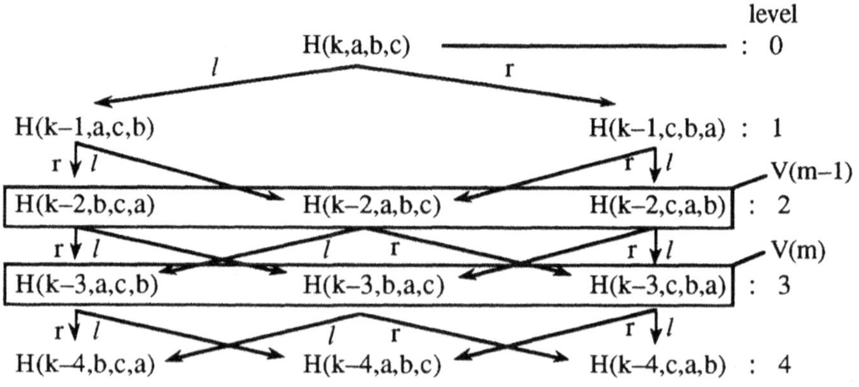


Figure 17: Reduced s-graph from level 0 to level 4 for the function:
H(k,a,b,c) = **if** k=0 **then** skip **else** H(k–1,a,c,b) : ab : H(k–1,c,b,a).

Since Property (δ) holds at level m–1, that is, V(m–1) $_r$= V(m), we introduce the function (see level 2 and 3 of Fig. 17):

t(k,a,b,c) $=_{def}$ < H(k–2,a,b,c), H(k–2,b,c,a), H(k–2,c,a,b) >.

By expressing the components of function t(k+1,a,b,c) in terms of those of t(k,a,b,c), and adding the base case for k=2, we get:

$$
\begin{array}{lll}
7.2 & t(2,a,b,c) & = \langle \text{skip, skip, skip} \rangle \\
7.3 & t(k+1,a,b,c) & = \langle u : ab : v,\ w : bc : u,\ v : ca : w \rangle \\
 & & \quad \textbf{where } \langle u,v,w \rangle = t(k,a,c,b) \qquad \text{for } k \geq 2.
\end{array}
$$

Equation 7.3 is equal to 2.9 which we have derived in Example 3. It realizes the optimality goals stated at the beginning of the previous section.

We now express the function H(k,a,b,c) (see level 0 of Fig. 17) in terms of the function t(k,a,b,c) (see level 2 of Fig. 17). By adding the base cases and using the associativity of the concatenation operator ':', we get:

$$
\begin{array}{lll}
7.4 & H(0,a,b,c) & = \text{skip} \\
7.5 & H(1,a,b,c) & = ab \\
7.6 & H(k+2,a,b,c) & = H(k+1,a,c,b) : ab : H(k+1,c,b,a) = \\
 & & = u:ac:v:ab:w:cb:u \textbf{ where } \langle u,v,w \rangle = t(k+2,a,b,c) \quad \text{for } k \geq 0.
\end{array}
$$

The final program is made out of Equations 7.2 through 7.6. This program is linear recursive and in our model of computation it requires a linear number of processes.

Now we will perform a further transformation step by deriving an iterative program which requires a total amount of three memory cells only. Thus, if we assume that we need one process for the parallel updating of one cell, we need three computing processes only. Indeed, if we denote by $\lambda zxy.J(z,x,y)$ the function which interchanges the values of the pegs x and y in any expression or tuple z where they occur, we have that: t(k,a,c,b) = J(t(k,a,b,c),b,c), and J(J(z,x,y),x,y) = z. Thus, Equation 7.3 can be rewritten as:

$$
\begin{array}{lll}
7.3^{*} & t(k+1,a,b,c) & = \langle u : ab : v,\ w : bc : u,\ v : ca : w \rangle \\
 & & \quad \textbf{where } \langle u,v,w \rangle = J(t(k,a,b,c),b,c) \qquad \text{for } k \geq 2.
\end{array}
$$

If we use Equation 7.3*, instead of 7.3, during the evaluation of H(k,a,b,c) each call of the function t has the second, third, and forth argument equal to a, b, and c, respectively.

We can then transform Equations 7.2 and 7.3* into an iteration by using the program schema equivalence of Fig. 18, which can be proved by induction on K≥N.

*If* S(p,z,x,y) = if p then J(z,x,y) else z *and*
J(E,x,y) = E *and* J(J(z,x,y),x,y) = z *and*
J(R(u,v),x,y) = R(J(u,x,y), J(v,x,y)) *then*

```
                                          {k = K ≥ N}
                                res := E;  p := even(k);
 T(N, z) = E                ≡   while k>N do begin  res := R(S(p,z,x,y), res);
 T(k+1, z) = R(z, J(T(k,z),x,y))                     p := not(p);  k := k-1
                                          end
                                       {res = T(K, z)}
```

Figure 18: A schema equivalence from linear recursion to iteration.

The matching substitution is:

$\{N = 2,\ \ Z = \langle a,b,c \rangle,\ \ x = b,\ \ y = c,\ \ E = \langle \text{skip,skip,skip} \rangle,\ \ T = \lambda kabc.t(k,a,b,c),$

$R = \lambda\langle a,b,c\rangle s. \langle u : ab : v,\ \ w : bc : u,\ \ v : ca : w\rangle$ **where** $\langle u,v,w\rangle = s\}$,
where $a, b, c \in Peg$, $z \in Peg^3$, and $E, T, s \in (\{AB, BC, CA, BA, CB, AC\}^*)^3$.

Thus, we have that:

$R(S(p,z,b,c), res) = \langle u : S(p,ab,b,c) : v,\ \ w : S(p,bc,b,c) : u,\ \ v : S(p,ca,b,c) : w\rangle$ **where** $\langle u,v,w\rangle = res$.

We get the following program:

```
                    {k = K ≥ 0}
if k=0 then res := skip else
if k=1 then res := ab else  {k = K ≥ 2}
begin  res := < skip, skip, skip >; p := even(k);
while k>2 do begin  res := < res1 : S(p,ab,b,c) : res2,   res3 : S(p,bc,b,c) : res1,
                             res2 : S(p,ca,b,c) : res3 >;
                    p := not(p);  k := k-1
              end;      {res = t(K,a,b,c)}
res := res1 : ac : res2 : ab : res3 : cb : res1
end                     {res = H(K,a,b,c)}
```

where the assignment to res is a parallel assignment, and $res_j$ denotes the j-th projection of the tuple res, for j=1,2,3. Three processes only are needed for performing the parallel assignment, and since the recursive structure has been replaced by the iterative one, the whole computation of H(K,a,b,c) for any value of K can be performed using a total number of three processes only.

The above transformation from a linear recursive program to an iterative one can be performed also to the programs we have derived in the Examples 5 and 6 above. We leave this task to the interested reader. In particular, we can derive iterative programs for the function d(n) of Example 5 and the function f(i,h) of Example 6 which improve the ones presented in [9], because they require fewer memory cells. ■

# 7 Conclusions

We have presented a technique for the optimal parallel evaluation of a large class of functions defined by a recursive equation of the form: $f(x) =_{def}$ **if** p(x) **then** a(x) **else** b(c(x), f(*l*(x)), f(r(x))), using synchronized concurrent processes.

Our results can easily be extended to other kinds of equations which are straightforward generalizations of that one. For instance, one may consider the following definitions of the function f: i) $f(x) =_{def}$ **if** p(x) **then** a(x) **else** $b(c(x), f(h_1(x)), \ldots, f(h_n(x)))$ with n recursive calls, instead of two, or ii) $f(x) =_{def}$ **if** $p_1(x)$ **then** a(x) **else if** $p_2(x)$ **then** $b_1(c_1(x), f(l_1(x)), f(r_1(x)))$ **else** $b_2(c_2(x), f(l_2(x)), f(r_2(x)))$, with two conditionals, instead of one only. One may also consider the case where the arity of the function f is larger than 1.

Minimal synchronizations among processes can be established at compile time by applying the tupling strategy. These synchronizations do not increase the parallel computation time and make use of auxiliary tupled functions which transform non-linear recursive programs into linear recursive ones. In our model of computation this transformation allows us to evaluate the given functions using a linear number of computing processes, avoiding all repeated computations of identical recursive calls, without increasing the total parallel computation time. In most cases only a constant number of processes are actually required for the evaluation of the derived programs.

The procedure we have presented produces in some examples better programs than the ones known in the literature [9].Somewhat related work can be found in [20], where sequential programs are transformed into parallel ones by enforcing some synchronizations.

# 8 Acknowledgements

Many thanks to Robert Paige and John Reif for their kind invitation to take part in the Workshop on 'Parallel Algorithm Derivation and Program Transformation'. The workshop gave us the opportunity of deepening our understanding of the problems which were discussed there, and it also provided the necessary stimulus for writing this paper. The warm hospitality by Robert and his family made the visit to New York very enjoyable and relaxing.

The University of Rome Tor Vergata and the IASI Institute of the National Research Council of Italy provided the necessary computing facilities.

# References

[1] Aerts, K., and Van Besien, D.: 'Implementing the Loop Absorption and Generalization Strategies in Logic Programs' Report of Electronics Department, Rome University Tor Vergata, 1991.

[2] Augustsson, L. and Johnsson, T.: 'Parallel Graph Reduction with the <v,G>-machine' *Proceedings of Functional Programming Languages and Computer Architecture*, London, 1989, 202-213.

[3] Barendregt, H.P.: *The Lambda Calculus, its Syntax and Semantics*, North-Holland (Amsterdam) 1984.

[4] Barendregt, H.P., van Eekelen, M.C.J.D., Glauert, J.R.W., Kennaway, J.R., Plasmeijer, M.J., and Sleep, M.R.: 'Term Graph Rewriting' PARLE Conference, *Lecture Notes in Computer Science* n. 259, 1987, 141-158.

[5] Bird, R.S.: 'The Promotion and Accumulation Strategies in Transformational Programming', *ACM Transactions on Programming Languages and Systems*, Volume 6, No. 4, 1984, 487-504.

[6] Burstall, R.M. and Darlington, J.: 'A Transformation System for Developing Recursive Programs', *Journal of the ACM*, Volume 24, No. 1, 1977, 44-67.

[7] Bush, V. J., and Gurd, J. R.: 'Transforming Recursive Programs for Execution on Parallel Machines' Proceedings of Functional Programming Languages and Computer Architecture, Nancy, France, *Lecture Notes in Computer Science* n. 201, Springer Verlag, 1985, 350-367.

[8] CIP Language Group: 'The Munich Project CIP', *Lecture Notes in Computer Science* n. 183, Springer Verlag, 1985.

[9] Cohen, N. H.: 'Eliminating Redundant Recursive Calls' *ACM Transactions on Programming Languages and Systems*, Volume 5, 1983, 265-299.

[10] Courcelle, B.: 'Recursive Applicative Program Schemes', in *Handbook of Theoretical Computer Science*, Volume B, Chapter 9, Elsevier Science Publishers, 1990, 459-492.

[11] Darlington, J.: 'An Experimental Program Transformation' *Artificial Intelligence* 16, 1981, 1-46.

[12] Darlington, J. and Pull, H.: 'A Program Development Methodology Based on a Unified Approach to Execution and Transformation' *IFIP TC2 Working Conference on Partial and Mixed Computation*, Ebberup, Denmark (D. Bjørner and A. P. Ershov, editors), North Holland, 1987, 117-131.

[13] Darlington, J. and Reeve, M.: 'A Multi-Processor Reduction Machine for the Parallel Evaluation of Applicative Languages', *ACM Conference on Functional Programming Languages and Computer Architecture*, Portsmouth, New Hampshire, 1981, 65-75.

[14] Downey, P. J., Sethi, R. and Tarjan R. E.: 'Variations on the Common Subexpression Problem', *Journal of ACM*, Volume 27, No. 4, 1980, 758-771.

[15] Feather, M.S.: 'A System for Assisting Program Transformation', *ACM Transactions on Programming Languages and Systems*, 4 (1), 1982, 1-20.

[16] Feather, M.S.: 'A Survey and Classification of Some Program Transformation Techniques', Proceedings of the *TC2 IFIP Working Conference on Program Specification and Transformation*, Bad Tölz, Germany, 1986, 165-195.

[17] George, L.: 'An Abstract Machine for Parallel Graph Reduction' *Proceedings of Functional Programming Languages and Computer Architecture*, London, 1989, 214-229.

[18] Goldberg, B.: 'Buckwheat: Graph Reduction on a Shared-Memory Multiprocessor' *Proceedings of the ACM Conference on Lisp and Functional Programming*, 1988, 40-51.

[19] Gordon, M. J., Milner, R., and Wadsworth, C. P.: 'Edinburgh LCF', *Lecture Notes in Computer Science* n. 78, Springer Verlag, 1979.

[20] Janicki, R. and Müldner,T.: 'Transformation of Sequential Specifications into Concurrent Specifications by Synchronization Guards', *Theoretical Computer Science* 77, 1990, 97-129.

[21] Karp, R. M. and Ramachandran, V.: 'Parallel Algorithms for Shared-Memory Machines', *Handbook of Theoretical Computer Science*, 1990, 869-942.

[22] Kott, L: 'About Transformation System: A Theoretical Study', *3ème Colloque International sur la Programmation*, Dunod, Paris, 1978, 232-247.

[23] Landin, P. J.: 'The Mechanical Evaluation of Expressions' *Computer Journal* 6 (4), 1964, 308-320.

[24] Langendoen, K. G. and Vree, W. G.: 'FRATS: A Parallel Reduction Strategy for Shared Memory' Proceedings PLILP '91, *Lecture Notes in Computer Science* n. 528 (Maluszynski and Wirsing, editors), Springer Verlag, 1991, 99-110.

[25] Manna, Z.: *Mathematical Theory of Computation*, McGraw-Hill, 1974.

[26] Möller, B. (editor): 'Programs from Specifications', in Proceedings of the *IFIP TC2 Working Conference*, Asilomar Center, California, USA, North Holland (Amsterdam), 1991.

[27] Mosses, P. D.: 'Denotational Semantics' in *Handbook of Theoretical Computer Science*, Volume B, Chapter 9, Elsevier Science Publishers, 1990, 574-631.

[28] Paige, R. and Koenig, S.: 'Finite Differencing of Computable Expressions' *ACM Transactions on Programming Languages and Systems*, 4 (3), 1982, 402-454.

[29] Pettorossi, A.: 'Transformation of Programs and Use of Tupling Strategy', *Proceedings Informatica 77*, Bled, Yugoslavia, 1977, 3 103, 1-6.

[30] Pettorossi, A.: 'A Powerful Strategy for Deriving Efficient Programs by Transformation' *ACM Symposium on Lisp and Functional Programming*, Austin, Texas, USA, 6-8 August 1984, 273-281.

[31] Pettorossi, A. and Skowron, A.: 'Communicating Agents for Applicative Concurrent Programming', in Proceedings International Symposium on Programming, Turin, Italy, *Lecture Notes in Computer Science* n. 137 (Dezani-Ciancaglini and Montanari, editors), Springer Verlag, 1982, 305-322.

[32] Smith, D. R.: 'A Semiautomatic Program Development System' *IEEE Transactions on Software Engineering*, Volume 16, No. 9, 1990, 1024-1043.

[33] Staples, J.: 'Computation on Graph-like Expressions', *Theoretical Computer Science* 10, 1980, 171-185.

[34] Stoy, J. E.: *Denotational Semantics: The Scott-Scrachey Approach to Programming Language Theory*, The MIT Press, Cambridge, Massachusetts, 1977.

[35] Wadler, P. L.: 'Deforestation: Transforming Programs to Eliminate Trees', in Proceedings ESOP 88, Nancy, France, *Lecture Notes in Computer Science* n. 300, Springer Verlag, 1988, 344-358.

# 5

# Scheduling Program Task Graphs on MIMD Architectures

Apostolos Gerasoulis and Tao Yang
email: gerasoulis@cs.rutgers.edu, tyang@cs.rutgers.edu
Rutgers University, New Brunswick, New Jersey 08903, USA

**Abstract**

Scheduling is a mapping of parallel tasks onto a set of physical processors and a determination of the starting time of each task. In this paper, we discuss several static scheduling techniques used for distributed memory architectures. We also give an overview of a software system PYRROS [38] that uses the scheduling algorithms to generate parallel code for message passing architectures.

## 1 Introduction

In this paper we consider the scheduling problem for directed acyclic program task graphs (DAG). We emphasize algorithms for scheduling parallel architectures based on the asynchronous message passing paradigm for communication. Such architectures are becoming increasingly popular but programming them is very difficult since both the data and the program must be partitioned and distributed to the processors. The following problems are of major importance for distributed memory architectures:

1. The program and data partitioning and the identification of parallelism.

2. The mapping of the data and program onto an architecture.

3. The scheduling and co-ordination of the task execution.

From a theoretical point of view all problems above are extremely difficult in the sense that finding the optimum solution is NP-complete in general. In practice, however, parallel programs are written routinely for distributed memory architectures with excellent performance. Thus one of the grand challenges in parallel processing is if a compiler can be built that will *automatically* partition and parallelize a sequential program and then produce a schedule and generate the target code for a given architecture. For a specialized class of sequential program definitions, the identification of parallelism becomes simpler. For example, Peter Pepper in this book describes a methodology for identifying the parallelism in recursive program definitions. However, choosing good partitions even in this simple case is difficult and requires the computation of a schedule.

We present an overview of the scheduling problem. We emphasize static scheduling over dynamic, because we are interested in building an automatic scheduling and code generation tool with good performance for distributed memory architectures. Dynamic scheduling performs well for shared memory architectures with a small number of processors but not for distributed memory architectures. This is because dynamic scheduling suffers from high overhead at run-time. To fully utilize distributed memory architectures, the data and programs must be mapped at the "right" processors at compile time so that run time data and program movement is minimum. We have addressed the issues of static scheduling and developed algorithms along with a software system named PYRROS [38]. PYRROS takes as an input a task graph and produces schedules for message passing architectures such as nCUBE-II. The current PYRROS prototype has complexity "almost" linear in the size of the task graph and can handle task graphs with millions of tasks.

An automatic system for scheduling and code generation is useful in many ways. If the scheduling is determined at compile time then the architecture can be utilized better. Also a programmer does not have to get involved in low level programming and synchronization. The system can be used to determine a good program partitioning before actual execution. It can also be used as a testbed for comparing manually written scheduling with an automatically generated scheduling.

# 2 The Program Partitioning and Data Dependence

We start with definitions of the task computation model and architecture:

- **A directed acyclic weighted task graph** (DAG) is defined by a tuple $G = (V, E, \mathcal{C}, \mathcal{T})$ where $V = \{n_j, j = 1 : v\}$ is the set of task nodes and $v = |V|$ is the number of nodes, $E$ is the set of communication edges and $e = |E|$ is the number of edges, $\mathcal{C}$ is the set of edge communication costs and $\mathcal{T}$ is the set of node computation costs. The value $c_{i,j} \in \mathcal{C}$ is the communication cost incurred along the edge $e_{i,j} = (n_i, n_j) \in E$, which is zero if both nodes are mapped in the same processor. The value $\tau_i \in \mathcal{T}$ is the execution time of node $n_i \in V$.

- **A task** is an indivisible unit of computation which may be an assignment statement, a subroutine or even an entire program. We assume that tasks are convex, which means that once a task starts its execution it can run to completion without interrupting for communications, Sarkar [32].

- **The static macro-dataflow** model of execution, Sarkar [32], Wu and Gajski [35], El-Rewini and Lewis [9]. This is similar to the dataflow model. The data flow through the graph and a task waits to receive all data in parallel before it starts its execution. As soon as the task completes its execution it sends the output data to all successors in parallel.

## 2.1 Program partitioning

A program partitioning is a mapping of program statements onto a set of tasks. Since tasks operate on data, their input data must be *gathered* from a data structure and *transmitted* to the task before execution, then operated by the task and finally transmitted and *scattered* back to the data structure. If the data structure is distributed amongst many processors, then the *gather/scatter* and *transmission* operations are costly in terms of communication cost unless the data are *partitioned* properly. We present an example for partitioning a program.

The following simple program represents the Gaussian Elimination (GE) algorithm without pivoting. The data structure is an $n \times n$ two dimensional array.

```
GE kij form

for k = 1 : n - 1
  for i = k + 1 : n
    for j = k + 1 : n
      a(i, j) = a(i, j) - (a(i, k) * a(k, j))/a(k, k)

    end
  end
end
```

Figure 1: The *kij* form without pivoting for GE.

We first present a fine grain partitioning where tasks are defined at the statement level

$$u_{ij,k} : \{ \ a(i,j) = a(i,j) - a(i,k) * a(k,j)/a(k,k) \ \}.$$

This fine grain partitioning shown in Fig. 2(a) fully exposes the parallelism of the GE program but a fine grain machine architecture is required to exploit this parallelism. For coarse grain architectures, we need to use coarse grain program partitionings. Fig. 2(b) shows a coarse grain partitioning where the interior loop is taken as one task $U_k^i$. Each task $U_k^i$ modifies row $i$ using row $k$.

```
(a) kij fine grain partitioning

for k = 1 : n - 1
  for i = k + 1 : n
    for j = k + 1 : n
      [u_ij,k]
    end

  end
end
```

```
(b) kij coarse grain partitioning

for k = 1 : n - 1
  for i = k + 1 : n
  [U_k^i : { for j = k + 1 : n
              u_ij,k
           end} ]
  end
end
```

Figure 2: The *kij* - fine and coarse grain partitionings for GE.

## 2.2 Data dependence graph

Once a program is partitioned, data dependence analysis must be performed to determine the parallelism in the task graph. For $n = 4$ the fine and coarse grain dependence graphs corresponding to Fig. 2 are depicted in Fig. 3. The statement-level fine grain graph has the dependence edges between node $u_{ij,k}$ for $k = 1:3$ and $i, j = k+1:4$. Notice that task $u_{33,2}$ must begin execution after $u_{22,1}$ is completed since it uses the output of $u_{22,1}$. The direction of the dependence arrow shown in the DAG is determined by using the sequential execution of the *kij* program in Fig. 2. However, there is no dependence between $u_{22,1}$ and $u_{23,1}$ and they may be executed in parallel. All transitive edges have been removed from the graph.

The coarse grain graph is shown in Fig. 3 in ovals by aggregating several $u_{ij,k}$ into a coarser grain task $U_k^i$. We combine the edges between two oval tasks and a clear picture of the dependence task graph is shown as the U-DAG in Fig. 5.

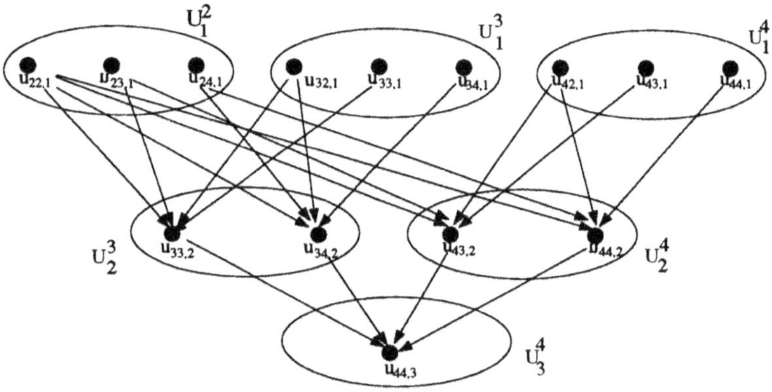


Figure 3: The fine grain DAG for GE and $n = 4$. Ovals show a coarse grain partitioning by aggregating small computation units $u_{ij,k}$.

## 2.3 Algorithms for partitioning

Partitioning algorithms need a cost function to determine if a partitioning is good or not. One widely used cost function is the minimization of the parallel time. Unfortunately, for this cost function the partitioning problem is NP-complete in most cases, Sarkar [32]. However, instead of searching for the optimum partitioning, we can search for a partitioning that has sufficient parallelism for the given architecture and also satisfies additional constraints. The additional constraints must be chosen so that the search space is reduced. An example of such a constraint is to search for tasks of a given maximum size that have no cycles. This is known as the *convexity constraint* in the literature, Sarkar [32]. A *convex* task is nonpreemptive in the sense that it receives all necessary data items before starting execution and completes its execution without any interruption. After that, it sends the data items to the successor tasks that need those data.

**Top-down:** One methodology for program partitioning is to start from the top level (the program) and go down the loop nesting levels until sufficient parallelism is discovered. At each loop level a partitioning is defined by mapping everything below that level in a task. Then a data dependence analysis is performed to find the parallelism at that level. If no sufficient parallelism is

```
for k = 1 : n − 1
  for j = k + 1 : n
   T_k^j : { for i = k + 1 : n
              u_ij,k
            end}
  end
end
```

Figure 4: The $kji$ - coarse grain partitioning for GE.

found at that level then program transformations such as loop interchange can be performed and test again the new loop for parallelism. Incorporating this loop interchange program transformation technique can also change the data access pattern of each task.

We show how the top-down approach works for the GE example. There are three nesting loop levels in the program of Fig. 1. Starting from the top (outer loop) we see that there is no parallelism. At the next level there is parallelism for some of the loops but the task convexity constraint sequentializes the task graph so we must go to the next level which is the interior loop level for our program. At the interior loop the tasks are convex and there is sufficient parallelism for coarse grain architectures as shown in the U-DAG in Fig. 5.

By loop interchanging loops $j$ and $i$ in the $kij$ GE program of Fig. 1 and taking the interior loop as a task, the result is the $kji$ form of GE algorithm shown in Fig. 4. The dependence graph is the T-DAG in Fig. 5. Each task $T_k^j$ uses column $k$ to modify column $j$.

**Bottom-up:** One difficulty with the top-down approach is that this approach follows the program structure level to partition and it is difficult to identify an appropriate level other than the statement level that has sufficient parallelism. Thus this approach will usually end up with a fine grain statement level task partitioning. If that is the case and we are interested in coarse grain partitioning then we must go bottom-up to determine such partitioning. Finding an optimal partitioning is NP-complete [32] and heuristics must be used.

We show an example of the bottom-up approach for Fig. 3. Given the fine grain DAG the partitioning in the ovals is a mapping corresponding to U-DAG coarse grain DAG. Another coarse grain partitioning is to aggregate $u_{22,1}$, $u_{32,1}$, $u_{42,1}$ into $T_1^2$ and so on; this results in the T-DAG shown in Fig. 5. The T-DAG and U-DAG have the same dependence structure but different task definitions. The two partitionings are also known as row and column partitionings because of their particular data access patterns.

## 2.4 Data partitioning

For shared memory architectures, the data structure is kept in a common shared memory while for distributed memory architectures, the data structure must be partitioned into *data units* and assigned to the local memories of the processors. A data unit can be a scalar variable, a vector or a submatrix block.

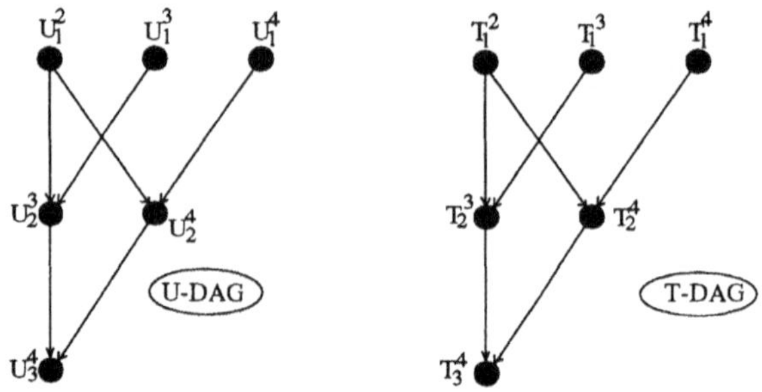


Figure 5: Dependence task graphs corresponding to two coarse grain partitioning. U-DAG with row data access pattern and T-DAG with column data access pattern

For distributed memory architectures large grain data partitioning is preferred because there is a high communication startup overhead in transferring a small size data unit. If a task requires to access a large number of distinct data units and data units are evenly distributed among processors, then there will be substantial communication overhead in fetching a large number of non-local data items for executing such task. Thus the following property can be used to determine program and data partitionings:

> **Consistency.** The program partitioning and data partitioning are consistent if sufficient parallelism is provided and also the number of distinct units accessed by each task is minimized.

Let us assume that the fine grain task graph in Fig. 3 is given and also that the data unit is a row of the matrix. Then the program partitioning shown in ovals is consistent with such a data partitioning and it corresponds to the U-DAG in Fig. 5. The resulting coarse grain tasks $U_k^j$ access an extensive number of data elements of rows $k$ and $j$ in each update. Making the data access pattern of a task consistent with data partitioning results in efficient reuse of data that reside in the local cache or the local memory.

Let us now assume that the matrix is partitioned in column data units. Then each task $U_k^j$ needs to access $n$ columns for each update, which results in excessive data movement. On the other hand, T-DAG task partitioning in Fig. 5 is consistent with column data partitioning since each task $T_k^j$ only accesses 2 columns ($k$ and $j$) for each update.

## 2.5 Computing the weights for the DAG.

Sarkar [32] on page 139 has proposed a methodology for the estimation of the communication and computation cost for the macro dataflow task model. The computation cost is the time $E$ for a task to execute on a processor. The communication cost consists of two components:

1. *Processor component:* The time that a processor participates in communication. The cost is expressed by the reading and writing functions $R$ and $W$.

2. *Transmission delay component:* The time $D$ for the transmission of the data between processors. During that time the processors are free to execute other instructions.

The weights can be obtained from

$$\tau_i = E_i, \qquad c_{ij} = R_i + D_{i,j} + W_j.$$

The parameters $R_i$, $D_{i,j}$, $W_j$ are functions of the *message size*, the *network load* and the *distance* between the processors. When there is no network contention, a very common approximation to $c_{i,j}$ is the *linear model*:

$$c_{ij} = (\alpha + k\beta)d(i,j)$$

where $\alpha$ is known as the startup time, $\beta$ is the transmission rate and $k$ is the size of the message transmitted between tasks $n_i$ and $n_j$ and $d(i,j)$ is the processor distance between tasks $n_i$ and $n_j$. This linear communication model is a good approximation to most currently available message passing architectures, see Dunigan [8]. For the nCUBE-II hypercube we have $\alpha = 160\mu s$ and $\beta = 2.4\mu s$ per word transfer for single precision arithmetic.

For the GE example, if $\omega$ is the time that it takes for each $u_{ij,k}$ operation, then $\tau_{kj} = (n-k)\omega$ for task $T_k^j$ in the T-DAG (or $U_k^i$ in the U-DAG) of Fig. 5. The communication weights are all equal to $(\alpha + (n-k)\beta)d(T_k^j, T_{k+1}^j)$, since only $(n-k)$ elements of the data unit are modified in $T_{k+1}^j$.

Of course, for some task graphs the computation and communication weights or even the dependence structure can only be determined at run time. For such cases run-time scheduling techniques are useful. For example, Saltz et. al. [30] and Koelbel and Mehrotra [22] use such an approach for problems that are iterative in nature. The program dependence graph is deterministic and can be derived during the first iteration at *run-time* and then run-time scheduling optimizations can be applied for other iterations. The initial overhead of such *run-time compilation* is usually high but this cost is amortized over all iterations. The scheduling techniques discussed in this paper can be applied as long as the dependence task graph is deterministic either at compile-time or at run-time.

# 3 Granularity and the Impact of Partitioning on Scheduling

## 3.1 Scheduling and clustering definitions

**Scheduling** is defined by a processor assignment mapping, $PA(n_j)$, of the tasks onto the $p$ processors and by a starting times mapping, $ST(n_j)$, of all nodes onto the real positive numbers set. Fig. 6(a) shows a weighted DAG with all computation weights assumed to be equal to 1. Fig. 6(b) shows a processor assignment using 2 processors. Fig. 6(c) shows a *Gantt chart* of a schedule for this DAG. The Gantt chart completely describes the schedule since it defines both $PA(n_j)$ and $ST(n_j)$. The scheduling problem has been shown to be NP-complete for a general task graph in most cases, Sarkar [32], Chretienne [4] and Papadimitriou and Yannakakis [27].

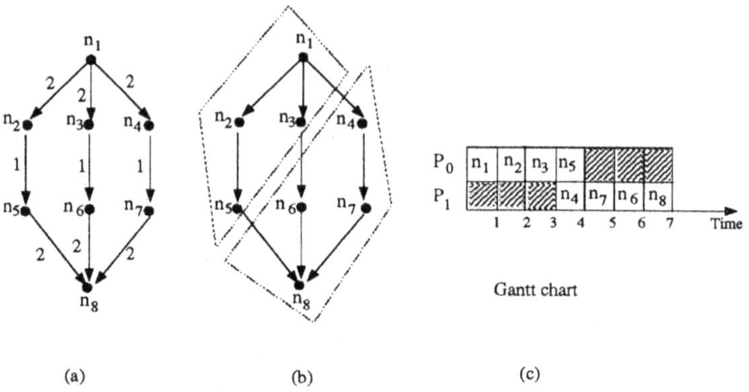


Figure 6: (a) A DAG with node weights equal to 1. (b) A processor assignment of nodes. (c) The Gantt chart of a schedule.

**Clustering** is a mapping of the *tasks* onto *clusters*. A *cluster* is a set of tasks which will execute on the same processor. Clusters are not tasks, since tasks that belong to a cluster are permitted to communicate with the tasks of other clusters immediately after completion of their execution. The clustering problem is identical to processor assignment part of scheduling in the case of an unbounded number of completely-connected processors. Sarkar [32] calls it an *internalization prepassing*. Clustering is also NP-complete for the minimization of the parallel time [4, 32].

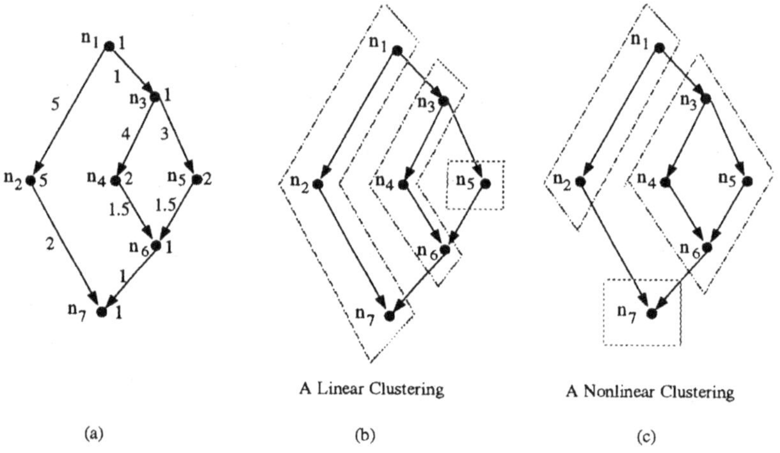


Figure 7: (a) A weighted DAG. (b) A linear clustering. (c) A nonlinear clustering.

A clustering is called *nonlinear* if two independent tasks are mapped in the same cluster, otherwise is called *linear*. In Fig. 7(a) we give a weighted DAG, in Fig. 7(b) a linear clustering with three clusters $\{n_1, n_2, n_7\}$, $\{n_3, n_4, n_6\}$, $\{n_5\}$ and in Fig. 7(c) a nonlinear clustering with clusters $\{n_1, n_2\}$, $\{n_3, n_4, n_5, n_6\}$ and $\{n_7\}$. Notice that for the nonlinear cluster independent tasks $n_4$ and $n_5$

are mapped in the same cluster.

In Fig. 8(a) we present the Gantt chart of a schedule for the nonlinear clustering of Fig. 7(c). Processor $P_0$ has tasks $n_1$ and $n_2$ with starting times $ST(n_1) = 0$ and $ST(n_2) = 1$. If we modify the clustered DAG as in [32] by adding a zero-weighted pseudo edge between any pair of nodes $n_x$ and $n_y$ in a cluster, if $n_y$ is executed immediately after $n_x$ and there is no data dependence edge between $n_x$ and $n_y$, then we obtain what we call a *scheduled DAG*. Fig. 8 (b) is a scheduled DAG and the dashed edge between $n_4$ and $n_5$ shows the pseudo execution edge.

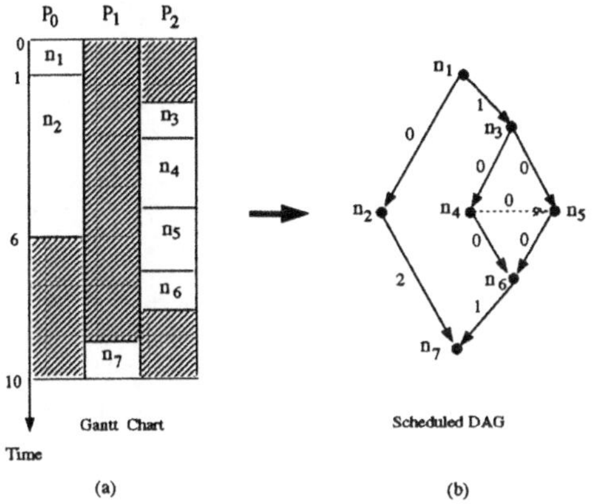


Figure 8: (a) The Gantt chart of a schedule for Fig. 1(c). (b) The scheduled DAG.

We call the longest path of the scheduled DAG the *dominant sequence* (DS) of the clustered DAG, to distinguish it from the *critical path* (CP) of a clustered but not scheduled DAG. For example, the clustered DAG in Fig. 7(c) has the sequence $< n_1, n_2, n_7 >$ as its CP with length 9, while a DS of this clustered DAG is $DS =< n_1, n_3, n_4, n_5, n_6, n_7 >$ and has length 10 using the schedule of Fig. 8(b). In the case of linear clustering, the DS and CP of the clustered DAG are identical, see Fig. 7(b).

## 3.2 The Granularity theory

One goal of partitioning is to produce a DAG that has sufficient parallelism for a given architecture. Another is to have a partition that minimizes the parallel time. These two goals are in conflict because having a partitioning with a high degree of parallelism does not necessarily imply the minimization of the parallel time, unless communication cost is zero. It is therefore the communication and computation costs derived by a partitioning that will determine the "useful parallelism " which minimizes the parallel time. This has been recognized in the literature as it can be seen by the following quote from Heath and Romine [19] p. 559:

> " Another important characteristic determining the overall efficiency of parallel algorithms is the relative cost of communication and computation. Thus, for example, if communication is relatively slow,

then coarse grain algorithms in which relatively large amount of computation is done between communications will be more efficient than fine-grain algorithms."

Let us consider the task graph in Fig. 9. If the computation cost $w$ is greater or equal to communication cost $c$ then the parallel time is minimum when $n_2$ and $n_3$ are executed in two separate processors as shown in Fig. 9(c). In this case all parallelism in this partitioned graph can be fully exploited since it is "useful parallelism". If on the other hand we assume that $w < c$, then the parallelism is not "useful" since the minimum parallel time is derived by sequentializing the tasks $n_2$ and $n_3$ as shown in Fig. 9(b).

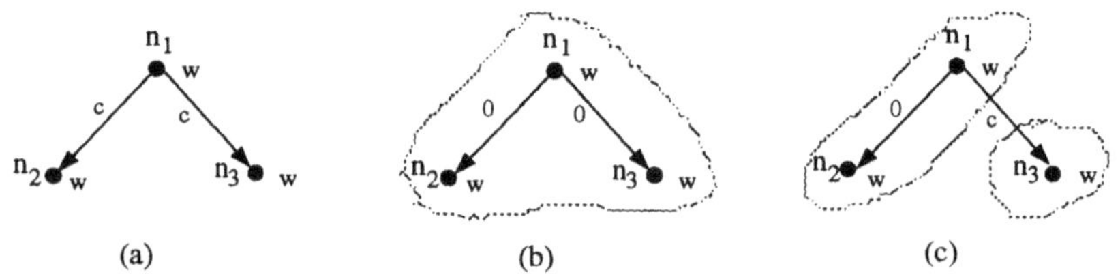


Figure 9: Sequentialization vs. parallelization. (a) A weighted DAG. (b) Sequentialization using nonlinear clustering. (c) Parallelization using a linear clustering.

Notice that linear clustering preserves the parallelism embedded in the DAG while nonlinear clustering does not. We make the following observation:

*If the execution of a DAG uses linear clustering and attains the optimal time, then this indicates that the program partitioning is appropriate for the given architecture; otherwise the partitioning is too fine and the scheduling algorithm still has to execute independent tasks together in the same processor using the nonlinear clustering strategy.*

It is therefore of interest to know when we can fully exploit the parallelism in a given task graph. We make the following assumption on the architecture:

The architecture is a clique with unbounded number of processors, i.e. a completely connected distributed memory architecture.

In Fig. 9 we saw the impact of the ratio $w/c$ on scheduling a simple DAG. An interesting question arises: can this analysis be generalized to arbitrary DAGs. In Gerasoulis and Yang [14] we have introduced a new notion of *granularity* using a ratio of the computation to communication costs taken over all fork and joins subgraphs of a task graph. The importance of this choice of granularity definition will become clear later on.

A DAG consists of *fork* or/and *join* sets such as the ones shown in figure 10. The join set $J_x$ consists of all immediate predecessors of node $n_x$. The fork set $F_x$ consists of all immediate successors of node $n_x$. Let $J_x = \{n_1, n_2, \ldots, n_m\}$ and $F_x = \{n_1, n_2, \ldots, n_m\}$ and define

$$g(J_x) = \min_{k=1:m} \{\tau_k\} / \max_{k=1:m} \{c_{k,x}\} \qquad g(F_x) = \min_{k=1:m} \{\tau_k\} / \max_{k=1:m} \{c_{x,k}\}.$$

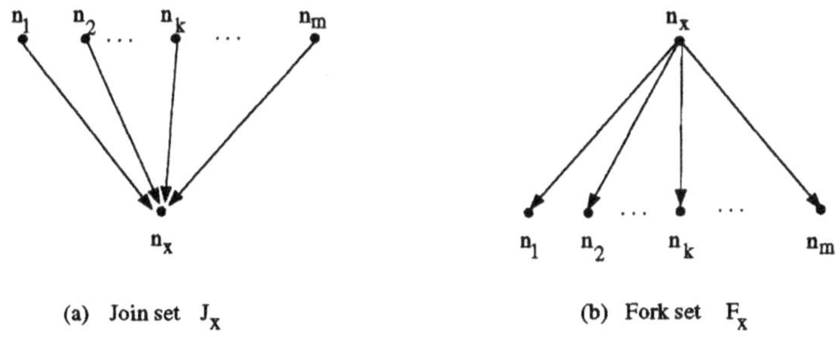


Figure 10: Fork and join sets.

We introduce the *grain* of a task $n_x$ as

$$g_x = \min\{g(F_x), g(J_x)\}$$

and the *granularity* of a DAG as

$$g(G) = \min_{x=1:v} \{g_x\}.$$

We call a DAG *coarse grain* if $g(G) \geq 1$ otherwise *fine grain*. If all task weights are equal to $R$ and all edge weights are equal to $C$ then the granularity reduces to $R/C$ which is the same as Stone's [34]. For coarse grain DAGs each task receives or sends data with a small amount of communication cost compared to the computation cost.

For example, the granularity of the graph in Fig. 7(a) is $g = 1/5$ which is derived as follows: The node $n_1$ is a fork and its *grain* is $g_1 = 1/5$, the ratio of the minimum computation weights of its successors $n_2$ and $n_3$, and the maximum communication cost of the outgoing edges. The node $n_2$ is in both a fork and a join and the grain for the join is $1/5$ which is the computation weight of its only predecessor $n_1$ and the cost of the edge $(n_1, n_2)$, while the grain for the fork is the weight of $n_7$ over the weight of $(n_2, n_7)$ which is again $1/2$. Continuing we finally determine the granularity as the minimum grain over all nodes of the graph which in our case is $g = 1/5$.

In [14] we prove the following theorems:

**Theorem 1** *For a coarse grain task graph, there exists a linear clustering that minimizes the parallel time.*

The above theorem is true only for our granularity definition and that is the reason for choosing it. We demonstrate the basic idea of the proof by using the example in Fig. 9. We show in [14] that for any nonlinear clustering we can extract a linear clustering whose parallel time is less than or equal to the nonlinear clustering. If we assume that $w \geq c$ in Fig. 9, then the parallel time of the nonlinear clustering in Fig. 9(b) is $3w$. By extracting $n_3$ from the nonlinear clustering and making it a new cluster, we derive a linear clustering shown in 9(c) whose parallel time is $2w + c \leq 3w$. We can always perform this extraction as long as the task graph is coarse grain.

Theorem 1 shows that the problem of finding an optimal solution for a coarse grain DAG is equivalent to that of finding an optimal linear clustering.

Picouleau [28] has shown that the scheduling problem for coarse grain DAGs is NP-complete, therefore optimal linear clustering is NP-complete.

**Theorem 2** *Determining the optimum linear clustering is NP-complete.*

Thus even though linear clustering is a nice property for task graphs, determining the optimum linear clustering is still a very difficult problem. Fortunately, for coarse grain DAGs, any linear clustering algorithm guarantees performance within a factor of two of the optimum as the following theorem demonstrates.

**Theorem 3** *For any linear clustering algorithm we have*

$$PT_{opt} \ \leq \ PT_{lc} \ \leq \ (1 + \frac{1}{g(G)})PT_{opt}$$

*where $PT_{opt}$ is the optimum parallel time and $PT_{lc}$ is the parallel time of the linear clustering. Moreover for a coarse grain DAG we have*

$$PT_{lc} \leq 2 \times PT_{opt}.$$

**Proof:** The proof is taken from [14]. Assume that the critical path is $CP = \{n_1, n_2, ..., n_k\}$. Then for any linear clustering, there could be some edges zeroed in that path but the length of that path $L_{cp}$ satisfies:

$$L_{cp} \leq \sum_{i=1}^{k} \tau_i + \sum_{i=1}^{k} c_{i,i+1}.$$

From the definition of the granularity we have that $g(G) \leq \tau_i / c_{i,i+1}$. Then by substituting $c_{i,i+1}$ in the last inequality we get

$$L_{cp} \leq \sum_{i=1}^{k} \tau_i (1 + \frac{1}{g(G)}).$$

Using the fact that

$$\sum_{i=1}^{k} \tau_i \leq PT_{opt} \leq PT_{lc} \leq L_{cp}$$

the inequality of the theorem is then derived easily. ■

Notice that when communication tends to zero then $g(G) \rightarrow +\infty$ and $PT_{opt} = PT_{lc}$. The above theorems provide an explanation of the advantages of linear clustering which has been widely used in the literature particularly for coarse grain dataflow graphs, e.g. [11, 23, 24, 26, 31]. We present an example.

**Example.** A widely used assumption for clustering is "the owner computes rule" [3], i.e. a processor executes a computation unit if this unit modifies the data that the processor owns. This rule can perform well for certain *regular* problems but in general it could result in workload imbalances especially for unstructured problems. The "owner compute rule" has been used to cluster both the U-DAG and the T-DAG in Fig. 5, see Saad [31], Geist and Heath [11] and Ortega [26]. This assumption results in the following clusters for the U-DAG shown in Fig. 11:

$$M_j = \{U_1^j, U_2^j, \ldots, U_k^j, \ldots, U_{j-1}^j\}, \quad j = 2:n$$

For each cluster $M_j$ row $j$ remains local in that cluster while it is modified by rows $1:j-1$ (similarly for columns in the T-DAG). The tasks in $M_j$ are chains in the task graph in Fig. 11 which imply that linear clustering was the result of the "owner computes rule". We call this special clustering, the *natural linear clustering.*

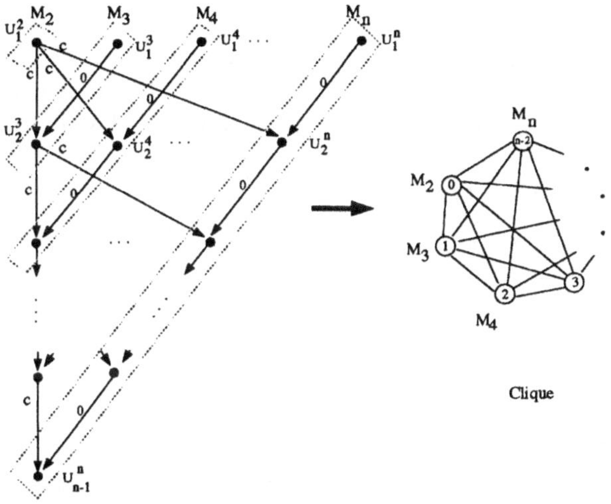


Figure 11: The natural linear clustering for the U-DAG executed on a clique with $p = n - 1$ processors

What is so interesting about the natural linear clustering? Let us assume that the computation size of all tasks is equal to $\tau$, and communication weights are equal to $c$ in the U-DAG. Then the following theorem holds:

**Theorem 4** *The natural linear clustering is optimal for executing the U-DAG on a clique architecture with* $(n-1)$ *processors provided the granularity* $g = \frac{\tau}{c} \geq 1$.

**Proof.** We can easily see that for $p = n - 1$ processors, the parallel time for the natural linear clustering is equal to the length of the critical path $\{U_1^2, U_2^3, \ldots, U_{n-1}^n\}$ of the scheduled U-DAG, see Fig. 11:

$$PT_{nlc} = (n-1)\tau \ + \ (n-2)c.$$

Since we have assumed that the granularity $g$ is greater than one, then Theorem 1 implies that *the optimum parallel time can be achieved by a linear clustering.* We have that

$$PT_{opt} \leq PT_{nlc} = (n-1)\tau \ + \ (n-2)c$$

and $PT_{opt}$ is the optimal parallel time using linear clustering. We will show that

$$PT_{opt} \geq (n-1)\tau \; + \; (n-2)c.$$

We define a *layer* $k$ of the U-DAG in Fig. 11 as the set of tasks $\{U_k^{k+1}, \ldots, U_k^n\}$. We will prove that the completion time of each task $U_k^j$ at layer $k$ satisfies

$$CT(U_k^j) \geq k\tau \; + \; (k-1)c, \; j = k+1 : n.$$

This is trivial for $k = 1$. Suppose it is true for tasks at layer $k-1$. We examine the completion time for each task $U_k^j$ at layer $k$. Since each task has two incoming edges from tasks at layer $k-1$, and linear clustering only zeros at most one edge, $U_k^j$ has to wait at least for time $c$ to receive the message from one of its two predecessors, say $U_{k-1}^r$, at layer $k-1$. Therefore

$$CT(U_k^j) \geq CT(U_{k-1}^r) + c + \tau.$$

From the induction hypothesis we have that

$$CT(U_{k-1}^r) \geq (k-1)\tau \; + \; (k-2)c$$

which implies

$$CT(U_k^j) \geq k\tau \; + \; (k-1)c.$$

Since the parallel time is the completion time of the last task $U_{n-1}^n$, this theorem holds. ∎

An application of Theorem 4 is the *kji* column partitioning form of Gauss-Jordan (GJ) algorithm. At each step of the GJ algorithm all $n$ elements of a column are modified and then transmitted to the successor tasks. The weights are then given by

$$\tau = n\omega, \qquad c = \alpha + n\beta$$

and as long as $n\omega/(\alpha+n\beta) \geq 1$, the GJ natural linear clustering is optimum. For the GE DAG, the weight of a task $U_k^i$ in U-DAG or $T_k^j$ in T-DAG is $(n-k)\omega$ and its incoming edge weights are $\alpha + (n-k)\beta$. For large $n$, only a small portion in the bottom of the DAG is fine grain and the natural clustering is asymptotically optimal by ignoring the insignificant low-order computation cost in this bottom portion.

We summarize our conclusions of this section as follows. For a program with coarse grain partitioning, linear clustering is sufficient to produce a good result. For a program with fine grain partitioning, linear clustering that preserves the parallelism of a DAG could lead to high communication overhead.

The granularity theory is a characterization of the relationship between partitioning and scheduling. In a real situation, some parts of a graph could be fine and others coarse. In such cases clustering and scheduling algorithm are needed to identify such parts and use proper clustering strategies to obtain the shortest parallel time. We consider these problems next.

# 4 Scheduling Algorithms for MIMD architectures

We distinguish between two classes of scheduling algorithms. The one step methods schedule a DAG directly on the $p$ processors. The multistep step methods perform a clustering step first, under the assumption that there are unlimited number of completely connected processors, and then in the following steps the clusters are merged and scheduled on the $p$ available processors. We consider heuristics that have the following properties: 1) They do not duplicate the same tasks in two different processors. 2) They do not backtrack.

## 4.1 One step scheduling methods

We present two methods. One is the classical list scheduling applied to the macro dataflow task graph and the other is the Modified Critical Path (MCP) heuristic proposed by Wu and Gajski [35].

**The classical list scheduling heuristic:**

The classical list scheduling schedules free[1] tasks by scanning a priority list from left to right. More specifically the following steps are performed:

1. Determine a priority list.
2. When a processor is *available* for execution, scan the list from left to right and schedule the first free task. If two processors are available at the same time, break the tie by scheduling the task in the processor with the smallest processor number.

When communication cost is zero, a good choice for a priority list is the Critical Path (CP) priority list. The priority of a task is its *bottom up level*, the length of the longest path from it to an exit node. The CP list scheduling possesses many nice properties when communication cost is zero. For example, it is optimum for tree DAGs with equal weights and for any arbitrary DAG with equal weights on 2 processors. For arbitrary DAGs and $p$ processors any list scheduling including CP is within 50% of the optimum. Moreover, the experimental results by Adam et. al. [1] show that CP is near optimum in practice in the sense that it is within 5% of the optimum in 90% of randomly generated DAGs. Unfortunately, these nice properties do not carry over to the case of nonzero communication cost.

In the presence of communication, it is extremely difficult to identify a good priority list. This is because the communication edge weight becomes zero when its end nodes are scheduled in the same processor and this makes the computation of the level priority information non-deterministic.

Let us consider the CP algorithm in the case where the level computation includes both edge communication and node computation. For example, a task graph is shown in Fig. 12(a) along with a list schedule based on the highest

[1] A task is *free* if all of its predecessors have completed execution. A task is *ready* if it is free and all of the data needed to start its execution is available locally in the processor where the task has been scheduled.

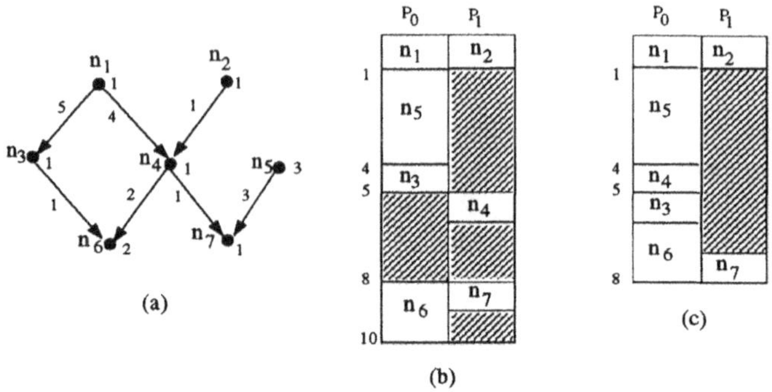


Figure 12: (a) A DAG. (b) The schedule by CP. (c) The schedule by MCP.

level first priority list. The level of $n_6$ is 2 and the level of $n_3$ is 4 which is equal to the maximum level of all successor tasks, which is 2, plus the communication cost in the edge $(n_3, n_6)$, which is 1, plus the computation cost of $n_3$, which is 1. The resulting priority list is $\{n_1, n_2, n_5, n_4, n_3, n_6, n_7\}$. Both $n_1$ and $n_2$ are free and the processors $P_0$ and $P_1$ available. At time 0, $n_1$ is scheduled in $P_0$ first and in the next step $n_2$ is scheduled in the only available processor $P_1$. At time 1, the tasks $n_3$, $n_4$ and $n_5$ are free and since $n_5$ has the highest priority it is scheduled in processor $P_0$ while the next highest priority $n_4$ is scheduled in processor $P_1$. Even if $n_4$ was scheduled in $P_1$ it needs to wait 4 unit times to receive the data from $P_0$ and thus $n_4$ is ready to start its execution at time 5. The task $n_5$ scheduled in $P_0$ can start execution immediately since the data are local in that processor. Continuing in a similar manner we get the final schedule shown in Fig. 12(b) with $PT = 10$.

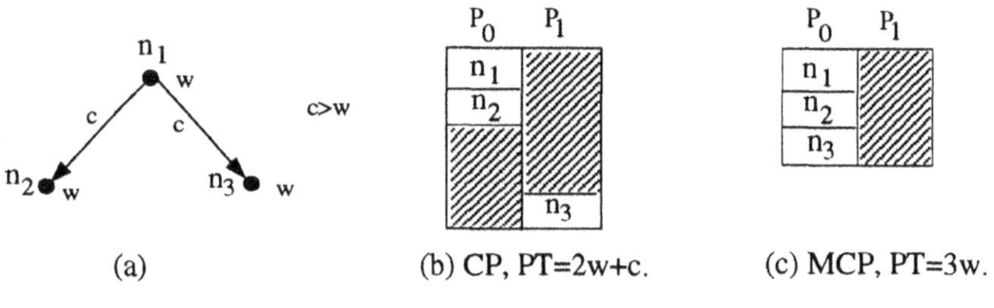


Figure 13: (a) A fork DAG. (b) The schedule by CP. (c) The schedule by MCP.

One problem with the CP heuristic in the presence of communication is that it schedules a *free* task when a processor becomes available, even though this task is not *ready* to start execution yet. This could result in poor performance as shown in Fig. 13(b). Task $n_3$ is scheduled in $P_1$ since it becomes free at time $w$. When $c > w$ a better solution is to schedule $n_3$ to $P_0$ shown in Fig. 13(c). We now present a modification to the CP heuristic.

**The modified critical path (MCP) heuristic:**

Wu and Gajski [35] have proposed a modification to the CP heuristic. Instead of scheduling a free task in an available processor, the free task is scheduled in the available processor that allows the task to start its execution at the *earliest* possible time. The computation of the priorities uses again the highest bottom up level including both communication and computation costs. For the example in Fig. 13(a), the priority list is $\{n_1, n_2, n_3\}$. The schedule is shown in Fig. 13(c). The task $n_3$ becomes free at time $w$ and it is scheduled in processor $P_0$ because it can start its execution at time $2w$ which is earlier than the time $w + c$ since $c > w$.

For the example in Fig. 12(a), the priority list is the same as in CP: $\{n_1, n_2, n_5, n_4, n_3, n_6, n_7\}$. After $n_1$, $n_2$ and $n_5$ are scheduled, task $n_4$ has the highest priority and is free at time 2 but is not ready at that time unless it is scheduled at $P_0$. Now $n_4$ is picked up for scheduling and it is scheduled in processor $P_0$ because it can start executing at time 4 which is earlier than time 5 if it was scheduled in $P_1$. The parallel time reduces to $PT = 8$ as depicted in Fig. 12(c).

Even though the MCP performs better than CP, it could still perform poorly as can be seen in the scheduling of a join DAG shown in Fig. 14. MCP gives the same schedule as CP and if the communication cost is greater than computation cost the optimum schedule executes all tasks in one processor. The MCP cannot recognize this since it uses the earliest starting time principle and it starts both $n_2$ and $n_3$ at time 0. One weakness of such one-pass scheduling is that the task priority information is non-deterministic because the communication cost between tasks will become zero if they are allocated in the same processor.

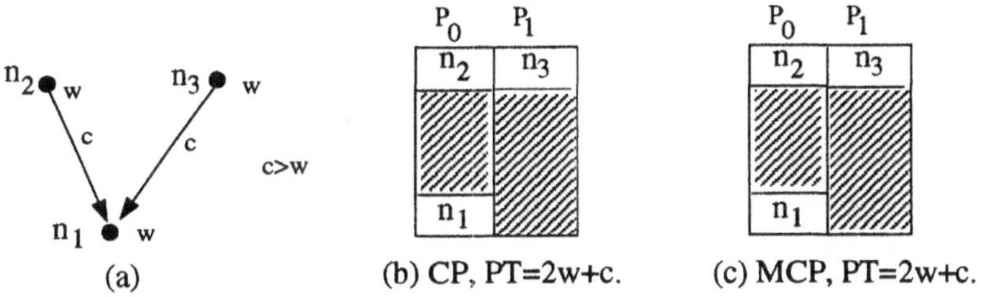


Figure 14: (a) A join DAG. (b) The schedule by CP. (c) The schedule by MCP.

It has been argued in the literature by Sarkar [32] and Kim and Browne [23] that a better approach to scheduling when communication is present is to perform scheduling in more than one steps. We discuss this approach next.

## 4.2 Multistep scheduling methods

**Sarkar's approach:**

Sarkar's heuristic [32] is based on the assumption that a scheduling pre-pass is needed to cluster tasks with high communication between them. Then the clusters are scheduled on $p$ available processors. To be more specific Sarkar advocates the following two step method:

1. Determine a clustering of the task graph by using scheduling on an unbounded number of processors and a clique architecture.

2. Schedule the clusters on the given architecture with a bounded number of processors.

Sarkar [32] uses the following heuristics for the two steps above:

1. Zero the edge with the highest communication cost. If the parallel time does not increase then accept this zeroing. Continue with the next highest edge until all edges have been visited.

2. After $u$ clusters are derived, schedule those clusters to $p$ processors by using a priority list. Assuming that the $v$ task nodes are sorted in a descending order of their priorities and the nodes are scanned from left to right. The scanned node, along with the cluster that it belongs to, is mapped on one of the $p$ processors that results in *the minimum increase in parallel time.* The parallel time is determined by executing the scheduled clusters in the physical processors and the unscheduled clusters in virtual processors.

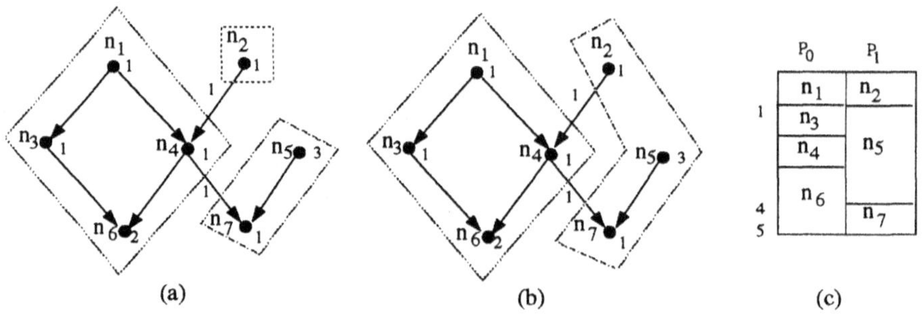


Figure 15: (a) The clustering result. (b) Clusters after merging. (c) The schedule with $PT = 5$.

Let us see how this two step method works for the example in Fig. 12(a). Initially the parallel time is 10. Sarkar's first clustering step zeroes the highest communication edge $(n_1, n_3)$ and the parallel time does not increase and this zeroing is accepted. The next highest edge $(n_1, n_4)$ is zeroed and the parallel time reduces, by executing $n_3$ either before or after $n_4$, so that this zeroing is also accepted. Next the edge $(n_5, n_7)$ is zeroed and after that the edge $(n_4, n_6)$ and the parallel time reduces to 5 which is determined by a $DS =< n_1, n_3, n_4, n_6 >$. By zeroing both $(n_2, n_4)$ or $(n_4, n_7)$ the parallel time increases and these zeroings are not accepted. The final result is three clusters:

$$M_1 = \{n_1, n_3, n_4, n_6\}, \quad M_2 = \{n_2\}, \quad \text{and } M_3 = \{n_5, n_7\}$$

shown in Fig. 15(a).

Assume there are two processor $P_0$ and $P_1$ available. The second step in Sarkar's' algorithm determines a priority list based on the highest level first principle. The initial list is $\{n_2, n_1, n_5, n_3, n_4, n_6, n_7\}$ because the level of $n_2$ is 5 while the level of $n_1$ is 4 and so on. The algorithm first picks $n_2$ to schedule and let us assume that it is scheduled in processor $P_1$. Next the task $n_1$ is chosen to be scheduled. If it is scheduled to $P_0$ then all nodes in $M_1$ are scheduled to

$P_0$ and $PT$ is 5. If it is scheduled to $P_1$ then $PT$ becomes 6 since now $n_1$ and $n_2$ must be sequentialized. Thus we assign $M_1$ to $P_0$. Next $n_5$ is scanned and it is scheduled to $P_0$ otherwise scheduling to $P_1$ will make $PT = 9$. Next $n_3$ is scanned, if it is assigned to $P_1$ then all other nodes in $M_1$ will be re-assigned to $P_1$ and $PT = 10$. Thus $n_3$ remains in $P_0$. Finally we have a schedule shown in Fig. 15(c).

**PYRROS's multistep scheduling algorithms:**

The PYRROS tool [38] uses a multistep approach to scheduling:

1. Perform clustering using the Dominant Sequence Algorithm (DSC).
2. Merge the $u$ clusters into $p$ completely connected virtual processors if $u > p$.
3. Map the $p$ virtual processors into $p$ physical processors.
4. Order the execution of tasks in each processor.

This approach has similarities to Sarkar's two step method. There is however a major difference. The algorithms used here are faster in terms of complexity. This is because we would like to test the multistep method on real applications and parallel architectures and higher complexity algorithms offer very little performance gains especially for coarse grain parallelism.

**The DSC clustering algorithm:**

Sarkar's clustering algorithm has a complexity of $O(e(v+e))$. Furthermore, zeroing the highest communication edges is not the best approach since this edge might not belong in the DS and as a result the parallel time cannot be reduced. In [36, 15] we have proposed a new clustering algorithm called the DSC algorithm which has been shown to outperform other algorithms from the literature, both in terms of complexity and parallel time. The DSC algorithm is based on the following heuristic:

- The parallel time is determined by the DS. Therefore if we want to reduce it we must zero at least one edge in the DS.
- A DS zeroing based algorithm could zero one or more edges in DS at a time. This zeroing can be done incrementally in a sequence of steps.
- A zeroing should be accepted if the parallel time reduces from one step to the next.

The DSC algorithm is a special case of a DS zeroing based algorithm that performs all steps in a time complexity "almost" linear in the size of the graph. We show how an algorithm based on DS zeroings works for the example of Fig. 12(a).

Fig. 16(a) is the initial clustering. The DS is shown in thick arrows. There are two dominant sequences in Fig. 16(a) with $PT = 10$. In the first step, the edge $(n_1, n_3)$ in one DS is zeroed as shown in Fig. 16(b). The new DS is $< n_1, n_4, n_6 >$ and $PT = 10$. This zeroing is accepted since PT does not

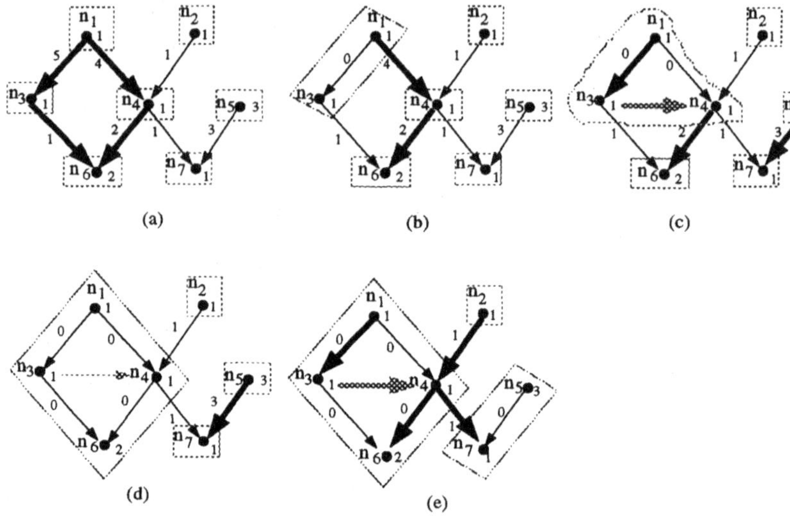

Figure 16: The clustering refinements in DSC.

increase. In the second step $(n_1, n_4)$ is zeroed and the result is two new DS $< n_1, n_3, n_4, n_6 >$ and $< n_5, n_7 >$ shown in Fig. 16(c) with $PT = 7$ and this zeroing is also accepted. In the third step $(n_4, n_6)$ is zeroed as shown in Fig. 16(d) and this zeroing is accepted since $PT = 7$ determined by DS $< n_5, n_7 >$. Next $(n_5, n_7)$ is zeroed and the PT is reduced to 5. Finally $(n_2, n_4)$ and $(n_4, n_7)$ cannot be zeroed because zeroing them will increase the parallel time. Thus three clusters are produced.

Notice that in the third step shown in Fig. 16(c) an ordering algorithm is needed to order the tasks in the nonlinear cluster and then the parallel time must be computed to get the new DS. One of the key ideas in the DSC algorithm is that it computes the schedule and parallel time incrementally from one step to the next in $O(\log v)$ time. Thus the total complexity is $O((v + e) \log v)$. If the parallel time was not computed incrementally, then the total cost would be greater than $O(v^2)$ which will not be practical for large graphs. More details can be found in [36].

**Cluster merging:**

The cost of the Sarkar cluster merging and scheduling algorithm is $O(pv(v + e))$ which is time-consuming for a large graph. PYRROS uses a variation of *work profiling method* suggested by George et. al. [16] for cluster merging. This method is simple and has been shown to work well in practice, e.g. Saad [31], Geist and Heath [11], Ortega [26], Gerasoulis and Nelken [12]. The complexity of this algorithm is $O(u \log u + v)$, which is less than $O(v \log v)$.

| |
|---|
| 1. Compute the arithmetic load $LM_j$ for each cluster. |
| 2. Sort the clusters in an increasing order of their loads. |
| 3. Use a load balancing algorithm so that each processor has approximately the same load. |

Let us consider an example. For the GE U-DAG in Fig. 11 there are $(n-1)$ clusters $M_2, M_3, \cdots, M_n$. We have that

$$LM_j = \sum_{i=1}^{j}(n-i)\omega \approx nj - \frac{j^2}{2}.$$

These clusters can be load balanced by using the wrap or reflection mapping, $VP(j) = (j-2) \bmod p$, Geist and Heath [11].

For the example in Fig. 15(a) with 3 clusters and 2 processors, the result of merging is two clusters shown in Fig. 15(b).

**Physical mapping:**

We now have $p$ virtual processors (or clusters) and $p$ physical processors. Since the physical processors are not completely connected, we must take the processor distance into account. Determining the optimum mapping of the virtual to physical processors is a very difficult problem since it can be instantiated as a Graph isomorphism problem.

Let us define $TC_{i,j}$ to be the total communication, which is the summation of costs of all edges between virtual processor $i$ and $j$. $CC = \{TC_{i,j} | TC_{i,j} \neq 0\}$ and $m = |CC|$. In general we expect that $m << e$.

The goal of the physical mapping is to determine the physical processor number $P(V_i)$ for each virtual processor $V_i$ that minimizes the following cost function F(CC):

$$F(CC) = \sum_{TC_{i,j} \in CC} distance(P(V_i), P(V_j)) \times TC_{i,j}.$$

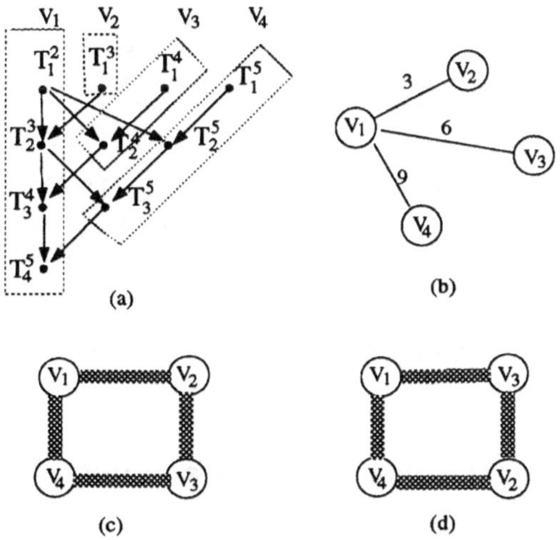


Figure 17: An example of physical mapping. Each nonzero edge cost is 3 time units. (a) A T-DAG linear clustering (b) Virtual cluster graph (c) A mapping to a hypercube (d) A better mapping.

Fig. 17 is an example of physical mapping for a T-DAG. A clustering for this DAG is shown in Fig. 17(a). The total communication between 4 virtual processors (clusters) is shown in Fig. 17(b). In Fig. 17(c) we show one physical mapping to a 4-node hypercube with $F(CC) = 24$ and another mapping is shown in (d) with $F(CC) = 21$.

Currently we use a heuristic algorithm due to Bokhari [2]. This algorithm starts from an initial assignment, then performs a series of pairwise interchanges so that the $F(CC)$ reduces monotonically as shown in the example above.

**Task ordering:**

Once the physical mapping has been decided then a task ordering is needed to define the scheduling. Since we no longer move tasks between processors the communication cost between tasks becomes deterministic. We show how important task ordering is via an example. The processor assignment along with communication and computation weights are shown in Fig. 18(a). In Fig. 18(b) we show one ordering with $PT = 12$ and in (c) another ordering in which the parallel time increases to $PT = 15$.

The task ordering that minimizes the parallel time is another NP-complete problem [10]. We have proposed a modification to the CP heuristic for the ordering problem in Yang and Gerasoulis [37]. This heuristic, Ready Critical Path (RCP), costs $O(v \log v + e)$ and is described below:

1. Adjust the communication edges of the DAG based on the processor assignment and physical distance.
2. Determine a global priority list based on the highest level first principle. The level computation includes both communication and computation cost in a path.
3. In addition to the global priority list each processor maintains a priority list of *ready* tasks for each processor. The ready task with the highest priority is executed as soon as this processor becomes free.

Let us consider the processor assignment in Fig. 18(a). The level priorities of tasks are: $L(n_1) = 12$, $L(n_2) = 7$, $L(n_3) = 1$, $L(n_4) = 1$, $L(n_5) = 2$, $L(n_6) = 2$. The priority list is $\{n_1, n_2, n_5, n_6, n_3, n_4\}$. Initially, $n_1$ is ready and is scheduled first on processor 0. At time 5, $n_2$ and $n_5$ are ready in processor 0 and $n_2$ is scheduled because of higher priority. The case is similar in processor 1 and $n_3$ is scheduled. The resulting schedule is shown in Fig. 18(b) and its parallel time is $PT = 12$.

## 4.3 Load balancing vs. Sarkar's cluster merging algorithms

As we discussed above PYRROS uses a simple heuristic based on load balancing for merging clusters. This heuristic uses only the cluster load information and completely ignores task precedences and inter-cluster communication. It is of interest to see how such a simple heuristic will perform vs. a more sophisticated,

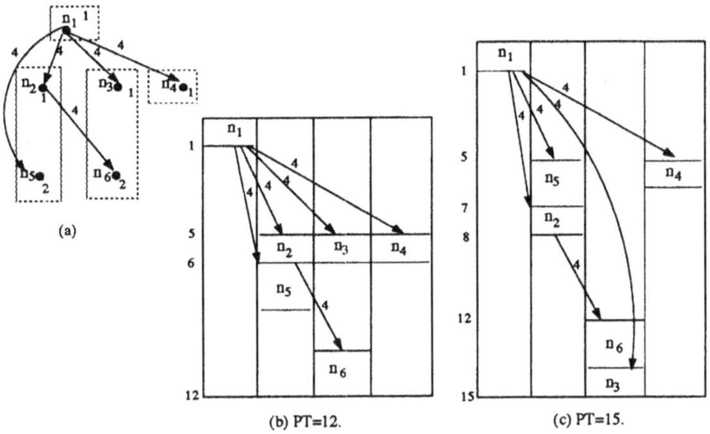


Figure 18: (a) A physical mapping of a DAG (b) The RCP ordering (c) Another ordering.

but more expensive in terms of complexity, heuristic such as Sarkar's cluster merging algorithm. To make a fair comparison, we use the same clustering algorithm for both cases, the DSC algorithm. We next merge the clusters using: (1) the load balancing heuristic (2) Sarkar's merging algorithm. We assume a clique architecture to avoid any mapping effects and use the RCP ordering in both cases to order tasks.

We generate randomly 100 DAGs and weights as follows: The number of tasks and edges are randomly generated and then assign randomly computation and communication weights. The size of the graphs varies from a minimum average of 143 nodes and 264 edges to a maximum average of 354 nodes and 2620 edges. In our experiments, the number of processors is chosen based on the widths of the graphs. The *width* and *depth* of graphs vary from 8 to 20 and thus we choose $p = 2, 4, 8$. Also to see the performance for both fine and coarse grain graphs we vary the granularity by varying the ratio of average computation over communication weights from 0.1 to 10.

Fig. 19 shows that the average improvement ratio (1-T(Sarkar)/T(Load balancing), where T() is the parallel time) of Sarkar's algorithm over the load balancing heuristic is between 10% to 35%. When the width of the graph is small compared to the number of processors, e.g. $p = 8$, Sarkar's algorithm is better than load balancing by about 30%. On the other hand, when the width is much larger than the number of processors then the performance differences are getting smaller, especially for coarse grain graphs, e.g. for $p = 2$ the improvement ratio reduces to about 10 % for coarse grain graphs. Intuitively, this is expected since each processor is assigned a larger number of tasks when the width to processor ratio increases and the RCP ordering heuristic can better overlap the computation and communication.

With respect to the execution time of the heuristics, for a Sun Sparcstation computer the load balancing heuristic takes about 0.1 seconds to produce a solution for graphs with average $v = 200$ and $e = 400$, while Sarkar's algorithm takes about 40 seconds. When we double the graph size, the load balancing

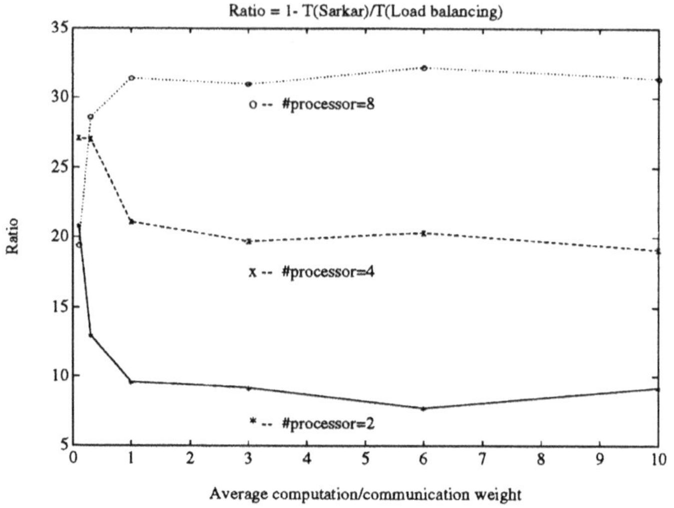


Figure 19: The performance of Sarkar's merging algorithm vs. load balancing algorithm. The graph width and depth are between 8 and 20.

heuristic takes 0.2 seconds while Sarkar's needs 160 seconds. For the above graphs and $p$, the time spent for each graph varied from 0.05 to 0.3 seconds for the load balancing heuristic and from 9.8 seconds to 725 seconds for Sarkar's. On the average, the load balancing heuristic was 1000 times faster than Sarkar's for those cases.

To verify our conclusions we increased the width of graphs from 8-20 to 30-40 but then reduced the depth of graphs between 5-8 to keep the number of tasks sufficiently small for the complexity of Sarkar's algorithm. The results are shown in Fig. 20 and are consistent with our previous conclusions. The performance of Sarkar's algorithm becomes better as the number of processors increases from $p = 2$ to $p = 16$ but then the performance reverses for $p = 32$, as expected, since $p$ approaches the width of the graph.

Our experiments show that on average the performance of the load balancing algorithm is within 75% of Sarkar's algorithm for those random graphs. This is very encouraging for the widely used load balancing heuristic. However, more experiments are needed to verify this result.

# 5 The PYRROS software tool

The input of PYRROS is a weighted task graph and the associated sequential C code. The output is a static schedule and parallel C code for a given architecture. The function modules of PYRROS are shown in Fig. 21. The current PYRROS tool has the following components: a task graph language with an interface to C, allowing users to define partitioned programs and data; a scheduling system for clustering the graph, load balancing and physical mapping, and communication/computation ordering; a graphic displayer for displaying task graphs and scheduling results; a code generator that inserts synchronization

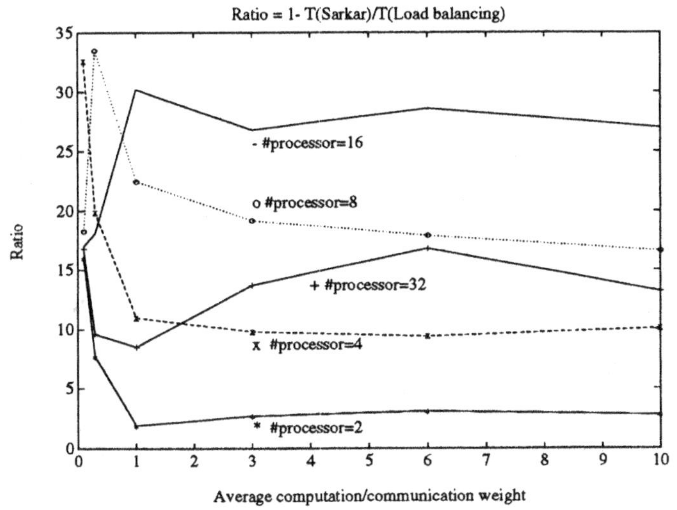


Figure 20: The performance of two merging algorithms for the graphs with width between 30 and 40 and depth between 5 and 8.

primitives and performs code optimization for nCUBE-I, nCUBE-II and INTEL iPSC/860 hypercube machines.

There are several other systems related to PYRROS. PARAFRASE-2 [29] by Polychronopoulos et. al., is a parallelizing compiler system that performs dependence analysis, partitioning and dynamic scheduling on shared memory machines. SCHEDULER by Dongarra and Sorensen [6] uses centralized dynamic scheduling for a shared memory machine. KALI by Koelbel and Mehrota [22] addresses code generation and is currently targeted at DOALL parallelism. Kennedy's group [20] is also working on code generation for FORTRAN D for distributed-memory machines. PARTI by Saltz's group [30] focuses on irregular dependence graphs determined at run-time and optimizes performance by precomputing data accessing patterns. HYPERTOOL by Wu and Gajski [35] and TASKGRAPHER by El-Rewini and Lewis [9] use the same task model as PYRROS. The time complexity of these two systems is over $O(v^2)$.

## 5.1 Task graph language

The PYRROS system uses a simple language for defining task graphs. For example, the program code in Fig. 22 is a description of the T-DAG partitioning shown in Fig. 4 and 5 in terms of PYRROS task graph language. The key words are boldfaced. The semantic of the loop is the same as that in Fig. 4. The interior loop body contains the data dependence and weight information for a task $T_k^j$ along with the specification of task computation. Task $T_k^j$ receives column $k$ and $j$ from $T_{k-1}^k$ and $T_{k-1}^j$ respectively if $k > 1$. The *c_update* is an external C function which defines the update of column $j$ using column $k$ operation of the task $T_k^j$ defined by the interior loop in the GE program in Fig. 4. After the *c_update* is executed, then if $k < n$ this task sends column $j$

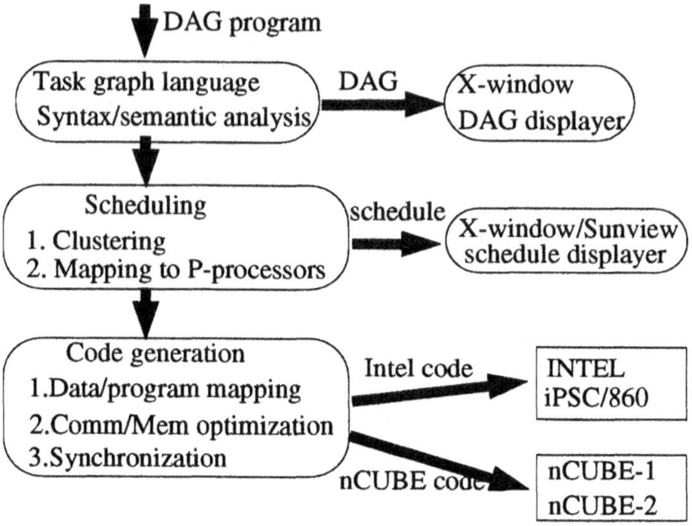


Figure 21: The system organization of PYRROS prototype.

to $T_{k+1}^{j}$ and also performs a broadcast to other tasks if $k = j - 1$.

PYRROS will read this program and perform lexical and semantic analysis to generate an internal representation of the DAG. Then using the X-window DAG displayer we can verify whether the definition of the task graph is correct.

## 5.2 A demonstration of PYRROS usage

In this section we demonstrate one usage of PYRROS. For GE T-DAG, we choose $\alpha = 10, \beta = \omega = 1, n = 5$ and PYRROS displays the dependence graph in the screen as shown in the left part of Fig. 23. Task $T(1,2)$ corresponds to task $T_1^2$ in the T-DAG of Fig. 5 and has an internal task number 1 written to its right. The edges of the DAG show the columns sent from one task to the successors.

As we mentioned above when a program is manually written for a library such as LINPACK [7], the clustering must be given in advance. Let us assume that the widely used natural linear clustering $M_j$ defined previously is used. This implies that $M_1 = \{T(1,2)\} = \{T1\}$, $M_2 = \{T(1,3), T(2,3)\} = \{T2, T5\}$ and so on. At this point the user, executing the program with natural linear clustering, cannot determine how many processors to choose so that the parallel time is minimized. If he chooses $p = 4$, because the width of the graph is 4 parallel tasks, the parallel time will be 75 time units shown in the right part of Fig. 23 after mapping clusters to processors. The striped lines in this Gantt chart represent communication delay on the hypercube with $p = 4$ processors. The internal numbers of tasks are used in the Gantt chart.

On the other hand, if the scheduling is determined automatically by PYRROS a better utilization of the architecture and shorter parallel time can be accomplished. In Fig. 24 PYRROS using the DSC algorithm determines that $p = 2$ processors are sufficient for scheduling this task graph and the parallel time is

```
for k=1 to n-1
   for j = k+1 to n
      task T(k,j)
            weight (n-k)*w
            receive if(k>1) column[k] T(k-1,k) weight alpha+beta*(n-k+1)
                  column[j] T(k-1,j) weight alpha+beta*(n-k+1)
               endif
           perform c_update(k,j)
           send if (k<n-1)
               if (k != j-1)
                  column[j] T(k+1,j) weight alpha+beta*(n-k)
               else
                  for b = j+1 to n
                     column[j] T(k+1,b) weight alpha+beta*(n-k)
                  endfor
               endif
         endif
      endtask
   endfor
endfor
```

Figure 22: PYRROS task specification for the T-DAG.

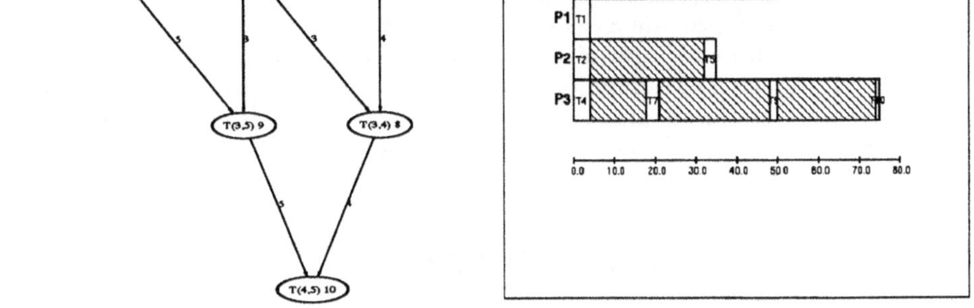


Figure 23: The left part is a GE DAG with $n = 5$ displayed in PYRROS X window screen. The right part is a Gantt chart using natural clustering.

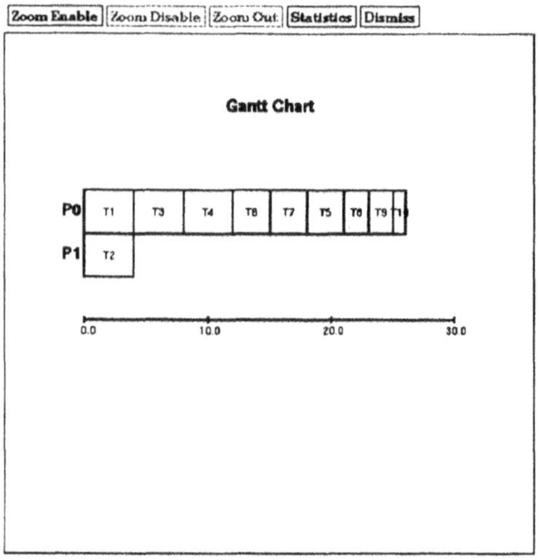


Figure 24: The automatic scheduling result by PYRROS.

reduced to 26 time units. The reason that natural clustering performs poorly here is that the graph is fine grain. Thus PYRROS is useful in determining the number of processors suitable for executing a task graph. This demonstrates one advantage of an automatic scheduling system.

## 5.3 Other PYRROS algorithms for code generation

The scheduling part of the current PYRROS prototype uses the previously-mentioned multi-step scheduling algorithms.

In addition to scheduling, PYRROS uses several other algorithms that generate code for message passing architectures such as INTEL and nCUBE. These algorithms distribute the data and program segments according to the processor assignment of tasks, insert communication primitives to achieve the correct synchronization and provide a deadlock-free communication protocol.

There are several code optimization techniques involved in code generation. One is the elimination of redundant interprocessor communications when there are duplicated data used among tasks. Another is the selection of efficient communication primitives based on the topology of target architectures. A more detailed description of each part of PYRROS system is given in [38].

## 5.4 Experiments with PYRROS

**The dense GE regular task graph computation:**

We report our experiments on the BLAS-3 [7] GE program in nCUBE-II. The dependence graph is similar to the one in Fig. 5 except that tasks operate on submatrices instead of array elements. The hand-written program uses the data column block partitioning with cyclic wrap mapping along the gray code

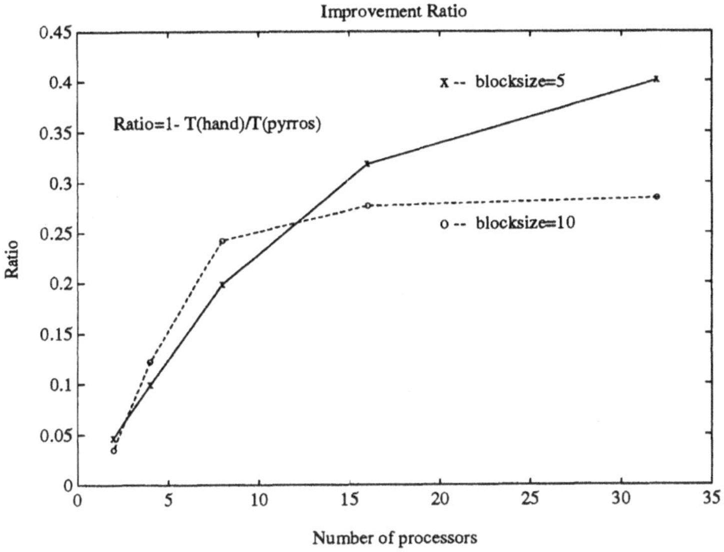


Figure 25: The improvement ratio of PYRROS over a hand-written GE program on nCUBE-II.

of a hypercube following the algorithm of Moler [25] and Saad [31]. Tasks that modify the same column block are mapped in the same processor. The broadcasting uses a function provided by the nCUBE-II library. The extra memory storage optimization for the hand-made program is not used to avoid the management overhead and as a consequence the maximum matrix size that this simple program can handle is $n = 450$.

The performance improvement of PYRROS code over this hand-written program, $Ratio = 1 - Time(hand)/Time(pyrros)$, for block sizes 5 and 10 is shown in Fig. 25. We can see the improvement is small for $p = 2$ because each processor has enough work to do. When $p$ increases, the PYRROS optimization plays an important role which results in 5% to 40% improvement. The speedup ratio of PYRROS over the sequential program for matrix size of 450 and 1000 is shown in the table below.

| | n=450 block size=5 | n=450 block size=10 | n=1000 block size=10 |
|---|---|---|---|
| p=2 | 1.97 | 1.9 | 1.99 |
| p=4 | 3.8 | 3.7 | 3.9 |
| p=8 | 7.3 | 6.9 | 7.8 |
| p=16 | 12.8 | 11.9 | 14.4 |
| p=32 | 19.0 | 12.9 | 25.7 |

**The sparse irregular task graph computation:**

One of the advantages of PYRROS is that it can handle irregular task graphs as well as regular ones. We present an example to demonstrate the importance of this feature.

Many problems in scientific computation involve sparse matrices. Generally dense matrix algorithms can be parallelized and vectorized fairly easily and attain good performance for most of the architectures. For example the LINPACK benchmark is a dense matrix benchmark. However if the matrices are sparse, the performance of the dense matrix algorithms is usually poor for most architectures. Thus the parallelization and efficient implementation of sparse matrix computations is one of the grand challenges in parallel processing.

An example is the area of circuit simulation and testing where the solution of a set of differential equations are often used. These problems are solved numerically by discretizing in time and space to reduce the problem to a large set of nonlinear equations. The solution is then obtained by an iterative method such as Newton-Raphson which iterates over the same dataflow graph derived in the first iteration, since the topology of the iteration matrix remains the same but the data change in each step. The following table shows the number of iterations of an iterative method over the same dataflow graph of several classes of problems taken from Karmarkar [21].

| LP Problems | Partial Differential Eqns | | Fractional Hypergraph Covering | | | Control Systems | | |
|---|---|---|---|---|---|---|---|---|
| Problem number | 1 | 2 | 1 | 2 | 3 | 1 | 2 | 3 |
| Repetition count of dataflow graph | 1428 | 12556 | 16059 | 6299 | 7592 | 708 | 1863 | 7254 |

The PYRROS algorithms could be very useful for this important class of problems. For the circuit simulation problem the LU decomposition method is used to determine the dataflow graph. Because many of the tasks in the graph perform updating with zeros, a naive approach is to modify the classical algorithms for finding LU so that the zero operations are skipped at run time. Such an approach is very inefficient because of the high overhead in operation skipping and also because of the difficulty in load balancing arithmetic. By traversing the task graph once we can delete all the zero nodes to derive the sparse irregular task graph. This traversal could be costly if it is done only once but this overhead is spread over many iterations and is usually insignificant, especially for symmetric matrices. There are other methods such as elimination tree algorithms that produce sparse graphs [17]. We show an example in Fig. 26. The left part is a dense LU graph with matrix size 9 while the right is a sparse LU graph after the deletion of useless tasks. PYRROS is perfect for such irregular problems since it can produce schedules that load balance arithmetic and also generate parallel code. Fig. 27 shows an experiment performed on nCUBE-II with matrix size 500. We compare code produced by PYRROS with a hand-made program that executes the dense graph using natural clustering *but skips zero operations.* The result shows that PYRROS code outperforms the hand-made regular program substantially. The reason is that it is difficult for a regular scheduling to perform well for an irregular graph.

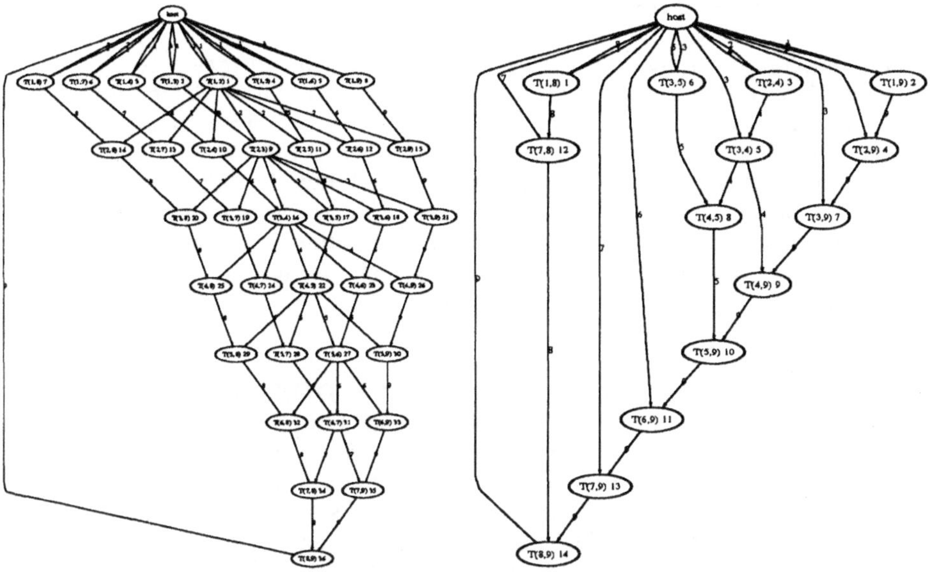


Figure 26: The left part is a dense LU DAG with $n = 9$. The right part is a sparse LU DAG with $n = 9$.

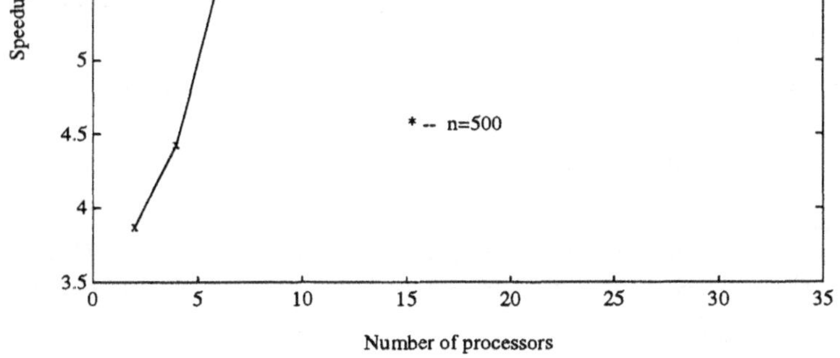


Figure 27: Dense with zero skipping vs. sparse dataflow graphs on nCUBE-II.

# 6 Conclusions

Scheduling program task graphs is an important optimization technique for scalable MIMD architectures. Our study on the granularity theory shows that scheduling needs to take communication overhead into account especially for message passing architectures. We have described several scheduling heuristic algorithms that attain good performance in solving the NP-hard scheduling problem. Those scheduling techniques are shown to be practical in PYRROS which integrates scheduling optimization with other compiler techniques to generate efficient parallel code for arbitrary task graphs.

## Acknowledgments

Partial support has been provided by a Grant No. DMS-8706122 from NSF and the Air Force Office of Scientific Research and the Office of Naval research under grant N00014-90-J-4018. We thank Weining Wang for developing the task graph language parser, Milind Deshpande for the X window schedule displayer, Probal Bhattacharjya for the graph generator of sparse matrix solver, and Ye Li for programming the INTEL i860 communication routines. We also thank referees and Ajay Bakre for their suggestions on the draft of this paper.

# References

[1] Adam, T., Chandy, K.M. and Dickson, J.R., 'A Comparison of List Schedules for Parallel Processing Systems', *CACM*, 17:12, 1974, pp. 685-690.

[2] Bokhari, S.H., '*Assignment Problems in Parallel and Distributed Computing*', Kluwer Academic Publisher, 1990.

[3] Callahan, D. and Kennedy, K., 'Compiling Programs for Distributed-memory Multi-processors', *Journal of Supercomputing*, Vol. 2, 1988, pp. 151-169.

[4] Chretienne, Ph., 'Task Scheduling over Distributed Memory Machines', *Proc. of Inter. Workshop on Parallel and Distributed Algorithms*, North Holland, 1989.

[5] Cosnard, M., Marrakchi, M., Robert, Y. and Trystram, D., 'Parallel Gaussian Elimination on an MIMD Computer', *Parallel Computing*, vol. 6, 1988, pp. 275-296.

[6] Dongarra, J.J. and Sorensen, D.C., 'SCHEDULE: Tools for Developing and Analyzing Parallel Fortran Programs', in *The Characteristics of Parallel Algorithms*, D.B. Gannon, L.H. Jamieson and R.J. Douglass (Eds), MIT Press, 1987, pp363-394.

[7] Dongarra, J.J., Duff, I., Sorensen, D.C. and van der Vorst, H.A., '*Solving Linear Systems on Vector and Shared Memory Computers*', SIAM,1991.

[8] Dunigan, T.H., 'Performance of the INTEL iPSC/860 and nCUBE 6400 Hypercube', ORNL/TM-11790, Oak Ridge National Lab., TN, 1991.

[9] El-Rewini, H. and Lewis, T.G., 'Scheduling Parallel Program Tasks onto Arbitrary Target Machines', *Journal of Parallel and Distributed Computing*, Vol. 9, 1990, pp. 138-153.

[10] Garey,M.R. and Johnson,D.S., '*Computers and Intractability: a Guide to the Theory of NP-completeness*', W.H. Freeman and Company (New York), 1979.

[11] Geist, G.A. and Heath,M.T., 'Matrix Factorization on a Hypercube Multiprocessor', *Hypercube Multiprocessors*, SIAM, 1986, pp. 161-180.

[12] Gerasoulis, A. and Nelken, I., 'Static Scheduling for Linear Algebra DAGs', Proc. of HCCA 4, 1989, pp. 671-674.

[13] Gerasoulis, A., Venugopal, S. and Yang, T., 'Clustering Task Graphs for Message Passing Architectures', *Proc. of 4th ACM Inter. Conf. on Supercomputing*, Amsterdam, 1990, pp. 447-456.

[14] Gerasoulis, A. and Yang, T., 'On the Granularity and Clustering of Directed Acyclic Task Graphs', TR-153, Dept. of Computer Science, Rutgers Univ., 1990.

[15] Gerasoulis, A. and Yang, T., 'A Comparison of Clustering Heuristics for Scheduling DAGs on Multiprocessors', To appear in *Journal of Parallel and Distributed Computing*, special issue on scheduling and load balancing, Dec. 1992.

[16] George, A., Heath, M.T., and Liu, J., 'Parallel Cholesky Factorization on a Shared Memory Processor', *Lin. Algebra Appl.*, Vol. 77, 1986, pp. 165-187.

[17] George, A., Heath,M.T., Liu, J. and Ng, E., 'Solution of Sparse Positive Definite Systems on a Hypercube', Report ORNL/TM-10865, Oak Ridge National Lab., 1988.

[18] Girkar, M. and Polychronopoulos, C., 'Partitioning Programs for Parallel Execution', *Proc. of ACM Inter. Conf. on Supercomputing*, St. Malo, France, 1988.

[19] Heath, M.T. and Romine,C.H., 'Parallel Solution of Triangular Systems on Distributed Memory Multiprocessors', *SIAM J. Sci. Statist. Comput.*, Vol. 9, 1988, pp. 558-588.

[20] Hiranandani, S., Kennedy, K. and Tseng,C.W., 'Compiler Optimizations for Fortran D on MIMD Distributed-Memory Machines', *Proc. of Supercomputing '91*, IEEE, pp. 86-100.

[21] Karmarkar, N., 'A New Parallel Architecture for Sparse Matrix Computation Based on Finite Project Geometries', *Proc. of Supercomputing '91*, IEEE, pp. 358-369.

[22] Koelbel, C., and Mehrotra, P., 'Supporting Shared Data Structures on Distributed Memory Architectures', *Proc. of ACM SIGPLAN Sympos. on Principles and Practice of Parallel Programming*, 1990, pp. 177-186.

[23] Kim, S.J. and Browne,J.C., 'A General Approach to Mapping of Parallel Computation upon Multiprocessor Architectures', *Proc. of Inter. Conf. on Parallel Processing*, Vol. 3, 1988, pp. 1-8.

[24] Kung, S.Y., *'VLSI Array Processors'*, Prentice Hall, 1988.

[25] Moler, C., 'Matrix Computation on Distributed Memory Multiprocessors', *Hypercube Multiprocessors 1986*, SIAM, pp. 181-195.

[26] Ortega, J.M., *'Introduction to Parallel and Vector Solution of Linear Systems'*, Plenum (New York), 1988.

[27] Papadimitriou, C. and Yannakakis,M., 'Towards on an Architecture-Independent Analysis of Parallel Algorithms', *SIAM J. Comput.*, Vol. 19, 1990, pp. 322-328.

[28] Picouleau, C., 'Two new NP-Complete Scheduling Problems with Communication Delays and Unlimited Number of Processors', M.A.S.I, Universite Pierre et Marie Curie Tour 45-46 B314, 4, place Jussieu, 75252 Paris Cedex 05, France, 1991.

[29] Polychronopoulos,C., Girkar, M., Haghighat, M., Lee,C., Leung, B., and Schouten, D., 'The Structure of Parafrase-2: an Advanced Parallelizing Compiler for C and Fortran',, in *Languages and Compilers for Parallel Computing*, D. Gelernter, A. Nicolau and D. Padua (Eds.), 1990.

[30] Saltz, J., Crowley, K., Mirchandaney, R. and Berryman,H., 'Run-Time Scheduling and Execution of Loops on Message Passing Machines', *Journal of Parallel and Distributed Computing,* Vol. 8, 1990, pp. 303-312.

[31] Saad, Y., 'Gaussian Elimination on Hypercubes', in *Parallel Algorithms and Architectures,* Cosnard, M. et al. (Eds.), Elsevier Science Publishers, North-Holland, 1986.

[32] Sarkar, V., *'Partitioning and Scheduling Parallel Programs for Execution on Multiprocessors'*, MIT Press, 1989.

[33] Sarkar, V., 'Determining Average Program Execution Times and their Variance', *Proc. of 1989 SIGPLAN*, ACM, pp. 298-312.

[34] Stone, H., *'High-Performance Computer Architectures'*, Addison-Wesley, 1987.

[35] Wu, M.Y. and Gajski, D., 'A Programming Aid for Hypercube Architectures', *Journal of Supercomputing*, Vol. 2, 1988, pp. 349-372.

[36] Yang, T. and A. Gerasoulis, A., 'A Fast Static Scheduling Algorithm for DAGs on an Unbounded Number of Processors', *Proc. of Supercomputing '91*, IEEE, pp. 633-642.

[37] Yang, T. and Gerasoulis, A., 'List Scheduling with and without Communication Delay', Report, 1992.

[38] Yang, T. and Gerasoulis, A., 'PYRROS: Static Task Scheduling and Code Generation for Message-Passing Multiprocessors', *Proc. of 6th ACM Inter. Confer. on Supercomputing,* Washington D.C., 1992, pp. 428-437.

# 6

# Derivation of Randomized Sorting and Selection Algorithms

**Sanguthevar Rajasekaran**

Dept. of CIS, University of Pennsylvania, raj@central.cis.upenn.edu

**John H. Reif**

Dept. of Computer Science, Duke University, reif@cs.duke.edu

**Abstract**

In this paper we systematically derive randomized algorithms (both sequential and parallel) for sorting and selection from basic principles and fundamental techniques like random sampling. We prove several sampling lemmas which will find independent applications. The new algorithms derived here are the most efficient known. From among other results, we have an efficient algorithm for sequential sorting.

The problem of sorting has attracted so much attention because of its vital importance. Sorting with as few comparisons as possible while keeping the storage size minimum is a long standing open problem. This problem is referred to as 'the minimum storage sorting' [10] in the literature. The previously best known minimum storage sorting algorithm is due to Frazer and McKellar [10]. The **expected** number of comparisons made by this algorithm is $n \log n + O(n \log\log n)$. The algorithm we derive in this paper makes only an expected $n \log n + O(n\, \omega(n))$ number of comparisons, for any function $\omega(n)$ that tends to infinity. A variant of this algorithm makes no more than $n \log n + O(n \log\log n)$ comparisons on **any input** of size $n$ with overwhelming probability.

We also prove high probability bounds for several randomized algorithms for which only expected bounds have been proven so far.

# 1 Introduction

## 1.1 Randomized Algorithms

A randomized algorithm is an algorithm that includes decision steps based on the outcomes of coin flips. The behavior of such a randomized algorithm is characterized as a random variable over the (probability) space of all possible outcomes for its coin flips.

More precisely, a randomized algorithm $A$ defines a mapping from an input domain $D$ to a set of probability distributions over some output domain $D'$. For each input $x \in D, A(x) : D' \to [0,1]$ is a probability distribution, where $A(x)\ (y) \in [0,1]$ is the probability of outputting $y$ given input $x$. In order for $A(x)$ to represent a probability distribution, we require

$$\sum_{y \in D'} A(x)(y) = 1, \text{for each } x \in D.$$

A mathematical semantics for randomized algorithms is given in [15].

Two different types of randomized algorithms can be found in the literature: 1) those which always output the correct answer but whose run time is a random variable; (these are called *Las Vegas* algorithms), and 2) those which output the correct answer with *high probability*; (these are called *Monte Carlo* algorithms). For example, the randomized sorting algorithm of Reischuk [27] is of the Las Vegas type and the primality testing algorithm of Rabin [19] is of the Monte Carlo type. In general, the use of probabilistic choice in algorithms to randomize them has often lead to great improvements in their efficiency. The randomized algorithms we derive in this paper will be of the Las Vegas type.

The amount of resource (like time, space, processors, etc.) used by a Las Vegas algorithm is a random variable over the space of coin flips. It is often difficult to compute the distribution function of this random variable. As an acceptable alternative people either 1) compute the expected amount of resource used (this bound is called the *expected bound*) or 2) show that the amount of resource used is no more than some specified quantity with 'overwhelming probability' (this bound is known as the *high probability bound*). It is always desirable to obtain high probability bounds for any Las Vegas algorithm, since such a bound provides a high confidence interval on the resource used. We say a Las Vegas algorithm has a resource bound of $\widetilde{O}(f(n))$ if there exists a constant $c$ such that the amount of resource used is no more than $c\alpha f(n)$ **on any input** of size $n$ with probability $\geq (1 - n^{-\alpha})$ (for any $\alpha > 0$). In an analogous manner, we could also define the functions $\widetilde{o}(.)$, $\widetilde{\Omega}(.)$, etc.

## 1.2 Comparison Problems and Parallel Machine Models

### *1.2.1 Comparison Problems*

Let $X$ be a set of $n$ distinct keys. Let $<$ be a total ordering over $X$. For each key $x \in X$ define

$$\underline{\text{rank}}(x, X) = |\{x' \in X | x' < x\}| + 1.$$

For each index $i, 1 \leq i \leq n$, we define $\underline{\text{select}}(i, X)$ to be that key $x \in X$ such that $i = \text{rank}(x, X)$. Also define

$$\underline{\text{sort}}(X) = (x_1, \ldots, x_n)$$

where $x_i = \text{select}(i, X)$, for $i = 1, \ldots, n$.

### *1.2.2 Parallel Comparison Tree Models*

In the *sequential comparison tree* model [16], any algorithm for solving a comparison problem (say sorting) is represented as a tree. Each non-leaf node in the tree corresponds to comparison of a pair of keys. Running of the algorithm starts from the root. We perform a comparison stored at the root. Depending on the outcome of this comparison, we branch to an appropriate child of the root. At this child also we perform a comparison and branch to a child, and so on. The execution stops when we reach a leaf, where the answer to the problem will be stored. The run time in this model is the number of nodes visited on a given execution. In a randomized comparison tree model execution from any node branches to a random child depending on the outcome of a coin tossing.

Valiant [31] describes a *parallel comparison tree* machine model which is similar to the sequential tree model, except that multiple comparisons are performed in each non-leaf of the tree. Thus a comparison tree machine with $p$ processors is allowed a maximum of $p$ comparisons at each node, which are executed simultaneously. We allow our parallel comparison tree machines to be randomized, with random choice nodes as described above.

### *1.2.3 Parallel RAM Models*

More refined machine models of computation also take into account storage and arithmetic steps. The sequential random access machine (RAM) described in [1] allows a finite number of register cells and also infinite global storage. A single step of the machine consists of an arithmetic operation, a comparison of two keys, reading off the contents of a global cell into a register, or writing the contents of a register into a global memory cell.

The parallel version of RAM proposed by Shiloach and Vishkin [29] (called the PRAM) is a collection of RAMs working in synchrony where communication takes place with the help of a common block of shared memory. For instance if processor $i$ wants to communicate with processor $j$ it can do so by writing a message in memory cell $j$ which then can be read by processor $j$.

Depending on whether concurrent reads and writes in the same memory cell by more than one processors are allowed or not, PRAMs can be further categorized into EREW (Exclusive Read and Exclusive Write) PRAMs, CREW (Concurrent Read and Exclusive Write) PRAMs, and CRCW PRAMs. In the case of CRCW, write conflicts can be resolved in many ways: On contention 1) an arbitrary processor succeeds, 2) the processor with the highest priority succeeds, etc.

### *1.2.4 Fixed Connection Networks*

These are supposed to be the most practical models. A number of machines like the MPP, connection m/c, $n$-cube, butterfly, etc. have been built based on these models. A fixed connection network is a directed graph whose nodes correspond to processing elements and whose edges correspond to communication links. Two processors which are connected by a link can communicate in a unit step. But if two processors which are not linked by an edge desire to communicate, they can do so by sending a message along a path that connects the two processors. Here again one could assume that each processor is a RAM. Examples include the mesh, hypercube, butterfly, CCC, star graph, etc.

The models we employ, in this paper, for various algorithms will be the ones used by the corresponding authors. We will explicitly state the models used.

## 1.3 Contents of this Paper

To start with, we derive and analyze a random sampling algorithm for approximating the rank of a key (in a set). This random sampling technique will serve as a building block for the selection and sorting algorithms we derive. We will analyze the run time for both the sequential and parallel execution of the derived algorithms.

The problem of selection also has attracted a lot of research effort. Many linear time sequential algorithms exist (see e.g., [1]). Reischuk's randomized selection algorithm [27] runs in $\widetilde{O}(1)$ time on the comparison tree model using

$n$ processors. Cole [8] has given an $O(\log n)$ [1] time $n/\log n$ CREW PRAM processor selection algorithm. Floyd and Rivest [11] give a sequential Las Vegas algorithm to find the $i$th smallest element in expected time $n + \min(i, n-i) + O(n^{2/3}\log n)$. We prove high probability bounds for this algorithm and also analyze its parallel implementation in this paper. The first optimal randomized network selection algorithm is due to Rajasekaran [22]. Followed by this work, several optimal randomized algorithms have been designed on the mesh and related networks (see e.g., [13, 21, 24]).

$\log(n!) \approx n\log n - n\log e$ is a lower bound for the comparison sorting of $n$ keys. Numerous **asymptotically** optimal sequential sorting algorithms like merge sort, heap sort, quick sort, etc. are known [16, 1]. Sorting with as few comparisons as possible while keeping the storage size minimum is an important problem. This problem is referred to as the minimum storage sorting problem. Binary merge sort makes only $n\log n$ comparisons but it needs close to $2n$ space to sort $n$ keys. A sorting algorithm that uses only $n + o(n)$ space is called a minimum storage sorting algorithm. The best known previous minimum storage sorting algorithm is due to Frazer and McKellar and this algorithm makes only an **expected** $n\log n + O(n\log\log n)$ number of comparisons. Remarkably, this expectation is over the space of coin flips. Even though this paper was published in 1970, this indeed is a randomized algorithm in the sense of Rabin [19] and Solovay & Strassen [30]. We present a minimum storage sorting algorithm that makes only $n\log n + \widetilde{O}(n\log\log n)$ comparisons. A variant of this algorithm needs only an expected $n\log n + O(n\ \omega(n))$ number of comparisons, for any function $\omega(n)$ that tends to infinity. Related works include: 1) A variant of Heapsort discovered by Carlsson [4] which makes only $(n+1)(\log(n+1) + \log\log(n+1) + 1.82) + O(\log n)$ comparisons in the worst case. (Our algorithms have the advantage of simplicity and less number of comparisons in the expected case); 2) Another variant of Heapsort that takes only an expected $n\log n + 0.67n + O(\log n)$ time to sort $n$ numbers [5]. (Here the expectation is over the space of all possible inputs, whereas in the analysis of our algorithms expectations are computed over the space of all possible outcomes for coin flips); and 3) Yet one more variant of Heapsort due to Wegener [32] that beats Quicksort when $n$ is large, and whose worst case run time is $1.5n\log n + O(n)$.

Many (asymptotically) optimal parallel comparison sorting algorithms are available in the literature. These algorithms are optimal in the sense that the product of time and processor bounds for these algorithms (asymptotically) equals the lower bound of the run time for sequential comparison sorting. These algorithms run in time $O(\log n)$ on any input of $n$ keys. Some of these algorithms are: 1) Reischuk's [27] randomized algorithm (on the PRAM model), 2) AKS deterministic algorithm [2] (on a sorting network based on expander graphs), 3) Column sorting algorithm due to Leighton [17] (which is an improvement in the processor bound of AKS algorithm), 4) FLASH SORT (randomized) algorithm of Reif and Valiant [25] (on the fixed connection network CCC), and 5) the deterministic parallel merge sort of Cole [7] (on the PRAM). On the other hand, there are networks for which no such algorithm can be designed. An example is the mesh for which the diameter itself is high (i.e., $2\sqrt{n} - 2$). Many optimal algorithms exist for sorting on the mesh and related networks

[1] All the logarithms mentioned in this paper are to the base 2, unless otherwise mentioned.

as well. See for example Kaklamanis, Krizanc, Narayanan, and Tsantilas [13], Rajasekaran [20], and Rajasekaran [21]. On the CRCW PRAM it is possible to sort in sub-logarithmic time. In [23], Rajasekaran and Reif present optimal randomized algorithms for sorting which run in time $\widetilde{O}(\frac{\log n}{\log\log n})$. In this paper we derive a nonrecursive version of Reischuk's algorithm on the CRCW PRAM.

In section 2 we prove several sampling lemmas which surely will find independent applications. One of the lemmas proven in this paper has been used to design approximate median finding algorithms [28]. In section 2 we also present and analyze an algorithm for computing the rank of a key approximately. In sections 3 and 4 we derive and analyze various randomized algorithms for selection and sorting. In section 5 our minimum storage sorting algorithm is given. Throughout this paper all samples are with replacement.

# 2 Random Sampling

## 2.1 Chernoff Bounds

The following facts about the tail ends of a binomial distribution with parameters $(n, p)$ will also be needed in our analysis of various algorithms.
Fact. *If $X$ is binomial with parameters $(n, p)$, and $m > np$ is an integer, then*

$$Probability(X \geq m) \;\leq\; \left(\frac{np}{m}\right)^m \; e^{m-np}. \tag{1}$$

*Also,*

$$Probability(X \leq \lfloor(1-\epsilon)pn\rfloor) \;\leq\; exp(-\epsilon^2 np/2) \tag{2}$$

*and*

$$Probability(X \geq \lceil(1+\epsilon)np\rceil) \;\leq\; exp(-\epsilon^2 np/3) \tag{3}$$

*for all $0 < \epsilon < 1$.*

## 2.2 An Algorithm for Computing Rank

Let $X$ be a set of $n$ keys with a total ordering $<$ defined on it. Our first goal is to derive an efficient algorithm to approximate rank$(x, X)$, for any key $x \in X$. We require that the output of our randomized algorithm have expectation rank$(x, X)$. The idea will be to sample a subset of size $s$ (where $s = o(n)$) from $X$, to compute the rank of $x$ in this sample, and then to infer its rank in $X$. The actual algorithm is given below.

```
algorithm samplerank_s(x, X);
    begin
        Let S be a random sample of X of size s;
        return ⌈1 + (n/s){rank(x, S) − 1})⌉
    end;
```

The correctness of the above algorithm is stated in the following

**Lemma 2.1** *The expected value of* samplerank$_s(x, X)$ *is rank(*$x, X$*).*

**Proof.** Let $k = \text{rank}(x, X)$. For a random $y \in X$, Prob.$[y < x] = \frac{k-1}{n}$. Hence, $E(\text{rank}(x, S)) = s\frac{k-1}{n} + 1$. Rewriting this we get

$$\text{rank}(x, X) = k = 1 + \frac{n}{s} E(\text{rank}(x, S) - 1) = E(\text{samplerank}_s(x, X)).\Box$$

Let $r_i = \text{rank}(\text{select}(i, S), X)$. The above lemma characterizes the expected value of $r_i$. In the next subsection we will obtain the distribution of $r_i$ using Chernoff bounds.

## 2.3 Distribution of $r_i$

Let $S = \{k_1, k_2, \ldots, k_s\}$ be a random sample from a set $X$ of cardinality $n$. Also let $k'_1, k'_2, \ldots, k'_s$ be the sorted order of this sample. If $r_i$ is the rank of $k'_i$ in $X$, the following lemma provides a high probability confidence interval for $r_i$.

**Lemma 2.2** *For every* $\alpha$, *Prob.* $\left(|r_i - i\frac{n}{s}| > c\alpha\frac{n}{\sqrt{s}}\sqrt{\log n}\right) < n^{-\alpha}$ *for some constant* $c$.

**Proof.** Let $Y$ be a fixed subset of $X$ of size $y$. We expect the number of samples in $S$ from $Y$ to be $y\frac{s}{n}$. In fact this number is a binomial $B(y, \frac{s}{n})$. Using Chernoff bounds (equation 3), this number is no more than $y\frac{s}{n} + \sqrt{3\alpha(ys/n)(\log_e n + 1)}$ with probability $\geq 1 - n^{-\alpha}/2$ (for any $\alpha$).

Now let $Y$ be the first $i\frac{n}{s} - \sqrt{3\alpha}\frac{n}{s}\sqrt{i(\log_e n + 1)}$ elements of $X$ in sorted order. The above fact implies that the probability that $Y$ will have $\geq i$ samples in $S$ is $\leq n^{-\alpha}/2$. This in turn means that $r_i$ is greater than or equal to $i\frac{n}{s} - \sqrt{3\alpha}\frac{n}{s}\sqrt{i(\log_e n + 1)}$ with probability $\geq 1 - n^{-\alpha}/2$.

Similarly one could show that $r_i$ is $\leq i\frac{n}{s} + \sqrt{2\alpha}\frac{n}{s}\sqrt{i(\log_e n + 1)}$ with probability $\geq (1 - n^{-\alpha}/2)$. Since $i \leq s$, the lemma follows. $\Box$

**Note:** The above lemma can also be proven from the fact that $r_i$ has a hypergeometric distribution and applying the Chernoff bounds for a hypergeometric distribution (derived in the appendix).

If $k'_1, k'_2, \ldots, k'_s$ are the elements of a random sample set $S$ in sorted order, then these elements divide the set $X$ into $(s + 1)$ subsets $X_1, \ldots, X_{s+1}$ where $X_1 = \{x \in X | x \leq k'_1\}$, $X_i = \{x \in X | k'_{i-1} < x \leq k'_i\}$, for $i = 2, \ldots, s$ and $X_{s+1} = \{x \in X | x > k'_s\}$. The following lemma provides a high probability upper bound on the maximum cardinality of these sets.

**Lemma 2.3** *A random sample* $S$ *of* $X$ *(with* $|S| = s$*) divides* $X$ *into* $s + 1$ *subsets as explained above. The maximum cardinality of any of the resulting subsets is* $\leq 2\frac{n}{s}(\alpha + 1)\log_e n$ *with probability greater than* $1 - n^{-\alpha}$. *(*$|X| = n$*).*

**Proof.** Partition the sorted $X$ into groups with $\ell$ successive elements in each group. That is, the first group consists of the $\ell$ smallest elements of $X$, the second group consists of the next $\ell$ elements of $X$ in sorted order, and so on. Probability that a specific group does not have a sample in $S$ is $= (1 - \frac{s}{n})^\ell$.

Thus the probability (call it $P$) that at least one of these groups does not have a sample in $S$ is $\leq n(1-\frac{s}{n})^{\ell}$. $P \leq n\, e^{(-\ell s/n)}$ (using the fact that $(1-\frac{1}{x})^{x} \leq \frac{1}{e}$ for any $x$). If we pick $\ell = \frac{n}{s}(\alpha+1)\log_e n$, $P$ becomes $\leq n^{-\alpha}$ for any $\alpha$. Thus the lemma follows. □

# 3 Derivation of Randomized Select Algorithms

## 3.1 A Summary of Select Algorithms

Let $X$ be a set of $n$ keys. We wish to derive efficient algorithms for finding select$(i, X)$ where $1 \leq i \leq n$. Recall we wish to get the correct answer always but the run time may be a random variable. We display a canonical algorithm for this problem and then show how select algorithms in the literature follow as special cases of this canonical algorithm. (The algorithms presented in this section are applicable not only to the parallel comparison tree model but also to the CREW PRAM model.)

```
algorithm canselect(i, X);
    begin
        select a bracket (i.e., a sample) B of X such that
        select(i, X) lies in this bracket with very high
        probability;
        Let i1 be the number of keys in X less than the
        smallest element in B;
        return canselect(i - i1, B)
    end;
```

Select algorithm of Hoare [12] chooses a random *splitter* key $k \in X$, and recursively considers either the low key set or the high key set based on where the $i$th element is located. And hence, $B$ for this algorithm is either $\{x \in X | x \leq k\}$ or $\{x \in X | x > k\}$ depending on which set contains the $i$th largest element of $X$. $|B|$ for this algorithm is $\frac{N}{c}$ for some constant $c$.

On the other hand, select algorithm of Floyd and Rivest [11] chooses two random splitters $k_1$ and $k_2$ and sets $B$ to be $\{x \in X | k_1 \leq x \leq k_2\}$. $k_1$ and $k_2$ are chosen properly so as to make $|B| = O(N^{\epsilon}), \epsilon < 1$. We'll analyze these two algorithms in more detail now.

## 3.2 Hoare's Algorithm

Detailed version of Hoare's select algorithm is given below.

```
algorithm Hselect(i, X);
    begin
        if X = {x} then return x;
        Choose a random splitter k ∈ X;
        Let B = {x ∈ X | x < k};
        if |B| ≥ i then return Hselect(i, B)
        else return Hselect(i - |B|, X - B)
```

end;

Let $\overline{T}_p(i,n)$ be the expected parallel time of Hselect$(i,X)$ using at most $p$ simultaneous comparisons at any time. Then the recursive definition of Hselect yields the following recurrence relation on $\overline{T}_p(i,n)$.

$$\overline{T}_p(i,n) = \frac{n}{p} + \frac{1}{n}\left[\sum_{j=1}^{i} \overline{T}_p(i-j, n-j) + \sum_{j=i+1}^{n} \overline{T}_p(i,j)\right].$$

An induction argument shows

$$\overline{T}_n(i,n) = O(\log n)$$

and

$$\overline{T}_1(i,n) \leq 2n + \min(i, n-i) + o(n)$$

To improve this Hselect algorithm, we can choose $k$ such that $B$ and $X - B$ are of approximately the same cardinality. This choice of $k$ can be made by fusing samplerank$_s$ into Hselect as follows.

```
algorithm sampleselect_s(i, X);
    begin
        if X = {x} then return x;
        Choose a random sample set S ⊆ X of size s;
        Let k = select(⌊s/2⌋, S);
        Let B = {x ∈ X | x < k};
        if |B| ≥ i then return sampleselect_s(i, B)
        else return sampleselect_s(i − |B|, X − B)
    end;
```

This algorithm can easily be analyzed using lemma 2.2.

## 3.3 Algorithm of Floyd and Rivest

As was stated earlier, this algorithm chooses two keys $k_1$ and $k_2$ from $X$ at random to make the size of its bracket $B = O(n^\beta), \beta < 1$. The actual algorithm is

```
algorithm FRselect(i, X);
    begin
        if X = {x} then return x;
        Choose k1, k2 ∈ X such that k1 < k2;
        Let r1 = rank(k1, X) and r2 = rank(k2, X);
        if r1 > i then FRselect(i, {x ∈ X | x < k1})
        else if r2 > i then FRselect(i − r1, {x ∈ X | k1 ≤
        x ≤ k2})
        else FRselect(i − r2, {x ∈ X | x > k2})
```

end;

Let $\overline{T}_p(i,n)$ be the expected run time of the algorithm FRselect$(i,X)$ allowing at most $p$ simultaneous comparisons at any time. Notice that we must choose $k_1$ and $k_2$ such that the case $r_1 \leq i \leq r_2$ occurs with high likelyhood and $r_2 - r_1$ is not too large. This is accomplished in FRselect as follows.

Choose a random sample $S \subseteq X$ of size $s$. Set $k_1$ to be select$\left(i\frac{s}{n} - \delta,\ S\right)$ and set $k_2$ to be select$\left(i\frac{s}{n} + \delta,\ S\right)$. If the parameter $\delta$ is fixed to be $\lceil d\alpha\sqrt{s\log n}\rceil$ for some constant $d$, then by lemma 2.2, Prob.$[r_1 > i] < n^{-\alpha}$ and Prob.$[r_2 < i] < n^{-\alpha}$. Let $\overline{T}_p(-,s) = \max_j \overline{T}_p(j,s)$. The resulting recurrence for the expected parallel run time with $p$ processors is

$$
\begin{aligned}
\overline{T}_p(i,n) &\leq \frac{n}{p} + \overline{T}_p(-,s) \\
&+\text{Prob.}[r_1 > i] \times \overline{T}_p(i,r_1) \\
&+\text{Prob.}[i > r_2] \times \overline{T}_p(i-r_1, n-r_2) \\
&+\text{Prob.}[r_1 \leq i \leq r_2] \times \overline{T}_p(i-r_1, r_2-r_1) \\
&\leq \frac{n}{p} + \overline{T}_p(-,s) + 2n^{-\alpha} \times n + \overline{T}_p\left(i, \left\lceil \frac{n}{s}\delta \right\rceil\right).
\end{aligned}
$$

Note that $k_1$ and $k_2$ are chosen recursively. If we fix $d\alpha = 3$ and choose $s = n^{2/3}\log n$, the above recurrence yields [11]

$$\overline{T}_1(i,n) \leq n + \min(i, n-i) + O(s).$$

Observe that if we have $n^2$ processors (on the parallel comparison tree model), we can solve the select problem in one time unit, since all pairs of keys can be compared in one step. This implies that $\overline{T}_p(i,n) = 1$ for $p \geq n^2$. Also, from the above recurrence relation,

$$\overline{T}_n(i,n) \leq O(1) + \overline{T}_n(-,\sqrt{n}) = O(1)$$

as is shown in [27].

## 3.4 High Probability Bounds

In the previous sections we have only shown expected time bounds for the selection algorithms. In fact only expected time bounds have been given originally by [12] and [11]. However, we can show that the same results hold with high probability. It is always desirable to give high probability bounds since it increases the confidence in the performance of the Las Vegas algorithms at hand.

To illustrate the method we show that Floyd and Rivest's algorithm can be modified to run sequentially in $n + \min(i, n-i) + \tilde{o}(n)$ comparison steps. This result may as well be a folklore by now (though to our knowledge it has not been published any where).

<u>algorithm</u> FR-Modified$(i, X)$;

<u>begin</u>

Randomly sample $s$ elements from $X$. Let $S$ be this sample;
Choose $k_1$ and $k_2$ from $S$ as stated in algorithm FRselect;
Partition $X$ into $X_1$, $X_2$, and $X_3$ where $X_1 = \{x \in X | x < k_1\}; X_2 = \{x \in X | k_1 \le x \le k_2\}$; and
$X_3 = \{x \in X | x > k_2\}$;
<u>if</u> select$(i, x)$ is in $X_2$ <u>then</u> deterministically compute and output select$(i - |X_1|, X_2)$ <u>else</u> start all over again

<u>end</u>;

**Analysis**. Since $s$ is chosen to be $n^{2/3}\log n$, both $k_1$ and $k_2$ can be determined in $O(n^{2/3}\log n)$ comparisons (using any of the linear time deterministic selection algorithms [1]). In accordance with lemma 2.2, the cardinality of $X_2$ will not exceed $c\alpha n^{2/3}$ with probability $\geq (1 - n^{-\alpha})$ (for some small constant $c$). Partitioning of $X$ into $X_1$, $X_2$, and $X_3$ can be accomplished with $n + \min(i, n-i) + \widetilde{O}(n^{2/3}\log n)$ comparisons using the following trick [11]: If $i \geq \frac{n}{2}$, always compare any key $x$ with $k_1$ first (to decide which of the three sets $X_1$ , $X_2$, and $X_3$ it belongs to), and compare $x$ with $k_2$ later only if there is a need. If $i < \frac{n}{2}$ do a symmetric comparison (i.e., compare any $x$ with $k_2$ first).

Given that select$(i, X)$ lies in $X_2$, this partitioning step can be performed within the stated number of comparisons. Also, selection in the set $X_2$ can be completed in $O(n^{2/3})$ steps.

Thus the whole algorithm makes only $n+\min(i, n-i)+\widetilde{O}(n^{2/3}\log n)$ number of comparisons. This bound can be improved to $n + \min(i, n-i) + \widetilde{O}(n^{1/2})$ using the 'improved algorithm' given in [11].

The same selection algorithm can be run on a CREW PRAM with a time bound of $\widetilde{O}(\log n)$ and a processor bound of $n/\log n$. This algorithm will then be an asymptotically optimal parallel algorithm. Along similar lines, one could also obtain optimal network selection algorithms [22, 13, 21, 24].

# 4 Derivation of Randomized Sorting Algorithms

## 4.1 A Canonical Sorting Algorithm

The problem is to sort a given set $X$ of $n$ distinct keys. The idea behind the canonical algorithm is to divide and conquer by splitting the given set into (say) $s_1$ disjoint subsets of almost equal cardinality, to sort each subset recursively, and finally to merge the resultant lists. A detailed statement of the algorithm follows.

<u>algorithm</u> cansort($X$)

<u>begin</u>

<u>if</u> $X = \{x\}$ <u>then</u> <u>return</u> $x$;
Choose a random sample $S$ from $X$ of size $s$;
Let $S_1$ be sorted $S$;
As explained in section 2.3, $S_1$ divides $X$
into $s+1$ subsets $X_1, X_2, \ldots, X_{s+1}$;
<u>return</u> cansort($X_1$).cansort($X_2$). $\cdots$. cansort($X_{s+1}$);

<u>end</u>;

Now we'll derive various sorting algorithms from the above.

## 4.2 Hoare's Sorting Algorithm

When $s = 1$ we get Hoare's algorithm. Hoare's sorting algorithm is very much similar to his select algorithm. Choose a random splitter $k \in X$ and recursively sort the set of keys $\{x \in X | x < k\}$ and $\{x \in X | x \geq k\}$.

<u>algorithm</u> quicksort($X$);

<u>begin</u>

<u>if</u> $|X| = 1$ <u>then</u> <u>return</u> $X$;
Choose a random $k \in X$;
<u>return</u> quicksort($\{x \in X | x < k\}$). $(k)$ . quicksort($\{x \in X | x > k\}$);

<u>end</u>;

Let $T_1(n)$ be the number of sequential steps required by quicksort($X$) if $|X| = n$. Then,

$$T_1(n) \leq n - 1 + \frac{1}{n} \sum_{1}^{n} (T_1(i-1) + T_1(n-i)) \leq 2n \log n.$$

A better choice for $k$ will be sampleselect$_s$($\lfloor n/2 \rfloor, n$). With this modification, quicksort becomes

<u>algorithm</u> samplesort$_s$($X$);

<u>begin</u>

<u>if</u> $|X| = 1$ <u>then</u> <u>return</u> $X$;
Choose a random sample $S$ from $X$ of size $s$;
Let $k$ =select($\lfloor s/2 \rfloor, S$);
<u>return</u> samplesort$_s$($\{x \in X | x < k\}$) . $(k)$ .
samplesort$_s$($\{x \in X | x > k\}$);

<u>end</u>;

By lemma 2.2,

$$\text{Prob.}\left[|\text{rank}(k, X) - n/2| > d\alpha \frac{n}{\sqrt{s}} \sqrt{\log n}\right] < n^{-\alpha}$$

for some constant $d$. If $\overline{C}(s, n)$ is the expected number of comparisons required by samplesort$_s(X)$, we have for $s(n) = n/\log n$,

$$\overline{C}(s(n), n) \leq 2\overline{C}(s(n_1), n_1) \ + \ n^{-\alpha}\overline{C}(s(n), n) \ + \ n \ + \ o(n)$$

where

$$n_1 = n/2 \ + \ d\alpha\sqrt{n} \log n.$$

Solving this recurrence Frazer and McKeller [10] show

$$\overline{C}(s(n), n) \approx n \log n,$$

which asymptotically approaches the optimal number of comparisons needed to sort $n$ numbers on the comparison tree model.

Let $\overline{\text{T}}_p(s, n)$ be the number of steps needed on a parallel comparison tree model with $p$ processors to execute samplesort$_s(X)$ where $|X| = n$. Since only a constant number of steps are required to select the median $k$ =select$(n/2, X)$ using $n$ processors, Reischuk [27] observes for this specialized algorithm with $s(n) = n$,

$$\begin{aligned}\overline{\text{T}}_n(n, n) &\leq O(1) + \overline{\text{T}}_{n/2}(n/2, n/2)\\ &= O(\log n).\end{aligned}$$

## 4.3 Multiple Sorting

Any algorithm with $s > 1$ falls under this category. Call cansort as multisort when $s > 1$. As was shown in Lemma 2.3, the maximum cardinality of any subset $X_i$ is $\leq 2(\alpha + 1)\frac{n}{s} \log_e n$ $(= n_1, \text{say})$ with probability $> 1 - O(n^{-\alpha})$. Therefore, if $\overline{\text{T}}_p(n)$ is the expected parallel comparison time for executing multisort$_s(X)$ with $p$ processors (where $|X| = n$) then,

$$\begin{aligned}\overline{\text{T}}_p(n) &\leq \overline{\text{T}}_{pn_1/n}(n_1) + n^{-\alpha}\overline{\text{T}}_p(n)\\ &\quad + \overline{\text{T}}_p(s) + \frac{n}{p} + \log(s)\\ &\leq \overline{\text{T}}_{pn_1/n}(n_1) + O(1) + \frac{n}{p}\log(s)\end{aligned}$$

Reischuk [27] uses the specialization $s = n^{1/2}$ which yields the following recurrence for $\overline{\text{T}}_p(n)$.

$$\overline{\text{T}}_n(n) \leq \overline{\text{T}}_{n_1}(n_1) + \frac{1}{2}\log n + O(1) = O(\log n)$$

Alternatively, as in [26], we can set $p = n^{1+\epsilon}$ and $s = n^{\epsilon}$ for any $0 < \epsilon < 1$ and get an $n_1 = n^{1-\epsilon/2} d\alpha\sqrt{\log n}$ for some constant $d$. This choice of $S$ yields the recurrence

$$\begin{aligned}\overline{\text{T}}_{n^{1+\epsilon}}(n) &\leq \overline{\text{T}}_{n^{\epsilon}n_1}(n_1) + O(1) + n^{-\epsilon}\log n\\ &= O(\log\log n)\end{aligned}$$

## 4.4 Non Recursive Reischuk's Algorithm

As stated above, Reischuk's algorithm is recursive. While it is easy to compute the expected time bound of a recursive Las Vegas algorithm, it is quite tedious to obtain high probability bounds (see e.g., [27]). In this section we modify Reischuk's algorithm so it becomes a non recursive algorithm. High probability bound of this modified algorithm will follow easily. This algorithm makes use of Preparata's [18] sorting scheme that uses $n \log n$ processors and runs in $O(\log n)$ time.

We assume a CRCW PRAM for the following algorithm.

**Step 1**

$s = n/(\log^4 n)$ processors randomly sample a key (each) from $X = k_1, k_2, \ldots, k_n$, the given input sequence.

**Step 2**

Sort the $s$ keys sampled in Step 1 using Preparata's algorithm. Let $l_1, l_2, \ldots, l_s$ be the sorted sequence.

**Step 3**

Let $X_1 = \{k \in X | k \leq l_1\}$; $X_i = \{k \in X | l_{i-1} < k \leq l_i\}$, $i = 2, 3, \ldots, s-1$; $X_s = \{k \in X | k > l_s\}$. Partition the given input $X$ into $X_i$'s as defined. This is done by first finding the part each key belongs to (using binary search in parallel). Now partitioning the keys reduces to sorting the keys according to their part numbers.

**Step 4**

For $1 \leq i \leq s$ in parallel do: sort $X_i$ using Preparata's algorithm.

**Step 5**

Output sorted$(X_1)$, sorted$(X_2)$,..., sorted$(X_s)$.

**Analysis.** Step 2 can be done using $s \log s$ ($\leq s \log n$) processors in $O(\log s)$ ($= O(\log n)$) time (see [18]).

In Step 3, binary search takes $O(\log n)$ time for each processor. Sorting the keys according to their part numbers can be performed in $\widetilde{O}(\log n)$ time and $n/\log n$ processors (see [23]), since this step is only sorting $n$ integers in the range $[1, s+1]$. Thus Step 3 can be performed in $\widetilde{O}(\log n)$ time, using $\leq n$ processors.

Using lemma 2.3, there will be no more than $O(\log^5 n)$ keys in each of the $X_i$'s ($1 \leq i \leq N$) with high probability. Within the same processor and time bounds, we can also count $|X_i|$ for each $i$. In Step 4, each $X_i$ can be sorted in $O(\log |X_i|)$ time using $|X_i| \log |X_i|$ processors. Also $X_i$ can be sorted in $(\log |X_i|)^2$ time using $|X_i|$ processors (using Brent's theorem). Thus Step 4 can be completed in $(\max_i \log |X_i|)^2$ time using $n$ processors. If $\max_i |X_i| = O(\log^5 n)$, Step 4 takes $O((\log \log n)^2)$ time. Thus we have proved the following

**Theorem 4.1** *We can sort n keys using n CRCW PRAM processors in $\tilde{O}(\log n)$ time.*

## 4.5 FLASHSORT

Reif and Valiant [25] give a method called FLASHSORT for dividing $X$ into even more equal sized subsets. This method is useful for sorts within fixed connection networks, where the processors can not be dynamically allocated to work on various size subsequences. The idea of Reif and Valiant [25] is to choose a subsequence $S \subset X$ of size $n^{1/2}$, and then choose as splitters every $(\alpha \log n)$th element of $S$ in sorted order, i.e., to choose $k_i'$ =select$(\alpha i \lfloor \log n \rfloor, S)$ for $i = 1, 2, \ldots, n^{1/2}/(\alpha \log n)$. Then they recursively sort each subset $X_i' = \{x \in X | k_{i-1}' < x \leq k_i'\}$. Their algorithm runs in time $O(\log n)$ and they have shown that after $O(\log n)$ recursive stages of their algorithm, the subsets will be of size no more than a factor of $O(1)$ of each other.

# 5 New Sorting Algorithm

In this section we present two minimum storage sorting algorithms. The first one makes only $n \log n + \tilde{O}(n \log \log n)$ comparisons, where as the second one makes an expected $n \log n + O(n \log \log \log n)$ number of comparisons. The second algorithm can be easily modified to improve the time bound further. The best known previous bound is $n \log n + O(n \log \log n)$ **expected number** of comparisons and is due to Frazer and McKellar [10].

The algorithm is similar to the one given in section 4.4. The only difference being that the sampling of keys is done in a different way. In section 4.4 $s = n/(\log n)^4$ keys were sampled at random from the input. On the other hand, here sampling is done as follows. 1) Pick a sample $S^*$ of $s'$ (for some $s'$ to be specified) keys at random from the input $X$; 2) Sort these $s'$ keys; 3) Keys in the sorted sequence in positions $1, (r+1), (2r+1), \ldots$ will belong to the sample (for some $r$ to be determined). In all, there will be $s = \lceil \frac{s'}{r} \rceil$ keys in the new sample (call it $S$). This sampling technique is similar to the one used by Reif and Valiant [25]. In fact, we generalize their sampling technique. We expect the new sample to 'split' the input more evenly.

Recall, if $s$ keys are randomly picked and each key is used as a splitter key, the input partition will be such that no part will be of size more than $\tilde{O}(\frac{n}{s} \log n)$. The new sampling will be such that no part will be of size more than $(1+\epsilon)\frac{n}{s}$, for some small $\epsilon$, with overwhelming probability ($s$ being the number of keys in the sample). We prove this fact before giving further details of our algorithm.

**Lemma 5.1** *If the input is partitioned using $s$ splitter keys (chosen in the manner described above), the cardinality of no part will exceed $(1+\epsilon)\frac{n}{s}$, with probability $\geq (1 - n^2\, e^{-\epsilon^2 n/s})$, for any $\epsilon > 0$.*

**Proof.** Let $x_0, x_1, \ldots, x_{f+1}$ be one of the longest ordered subsequences of sorted$(X)$ (where $f = (1+\epsilon)\frac{n}{s}$), such that $x_0, x_{f+1} \in S$ and $x_1, x_2, \ldots, x_f \notin S$.

The probability that out of the $s$ members of $S^*$, exactly $r$ lie in the above range and the rest outside is

$$\frac{\binom{n-f}{s'-r}\binom{f}{r}}{\binom{n}{s'}}.$$

The above is a hypergeometric distribution and as such is difficult to simplify. Another way of computing this probability is as follows. Each member of sorted($X$) is equally likely to be a member of $S^*$ with probability $\frac{s'}{n}$. We want to determine the length of a subsequence of sorted($X$) in which exactly $r$ elements have succeeded to be in $S$. This length is clearly the sum of $r$ identically distributed geometric variables each with a probability of success of $\frac{s'}{n}$. This has a mean of $\frac{rn}{s'} = \frac{n}{s}$. In the appendix we derive Chernoff bounds for the sum of geometric variables. Using this bound, Probability that $f \geq (1+\epsilon)\frac{n}{s}$ is $\leq e^{-\epsilon^2 n/s}$ (assuming $\epsilon$ is very small in comparison with 1).

There are at most $n^2$ choices for $x_0$ and $f$. Thus the lemma follows. □

## 5.1 An $n \log n + \tilde{O}(\log \log n)$ Time Algorithm

Frazer and McKellar's algorithm [10] for minimum storage sorting makes $n \log n + O(n \log \log n)$ expected number of comparisons. This expectation is over the coin flips. Even though this paper was published in 1970, the algorithm given is indeed a randomized algorithm in the sense of Rabin [19], and Solovay and Strassen [30]. Also, Frazer and McKellar's algorithm resembles Reischuk's algorithm [27]. In this section we present a simple algorithm whose time bound will match Frazer and McKellar's with overwhelming probability. The algorithm follows.

**Step 1**

Randomly choose a sample $S^*$ of $s' = n/\log^3 n$ keys from $X = k_1, k_2, \ldots, k_n$, the given input sequence. Sort $S^*$ and pick keys in positions $1, (r+1), \ldots$ where $r = \log n$. This constitutes the sample $S$ of $s = \lceil \frac{s'}{r} \rceil$ splitters.

**Step 2**

Partition $X$ into $X_i$, $1 \leq i \leq (s+1)$, using the splitter keys in $S$. (c.f. algorithm of section 4.4).

**Step 3**

Sort each $X_i$, $1 \leq i \leq (s+1)$ separately and output the sorted parts in the right order.

**Analysis**. Sorting in Step 1 and Step 3 can be done using any 'inefficient' $O(n \log n)$ algorithm. Thus, Step 1 can be completed in $O(n/(\log^2 n))$ time. Partitioning in Step 2 can be done using binary search on sorted($S$) and it takes $n(\log n - 4 \log \log n)$ comparisons. Using lemma 5.1, the size of no $X_i$ will be greater than $1.1 \log^4 n$ with overwhelming probability. Thus Step 3 can be finished in time $\sum_{i=1}^{s+1} O(|X_i| \log |X_i|) = O(n \log \log n)$.

Put together, the algorithm runs in time $n \log n + \tilde{O}(n \log \log n)$. □

## 5.2 An $n \log n + O(n\ \omega(n))$ Expected Time Algorithm

In this section we first modify the previous algorithm to achieve an expected time bound of $n \log n + O(n \log\log\log n)$. The modification is to perform one more level of recursion in the Reischuk's algorithm. Later we describe how to improve the time bound to $n \log n + O(n\ \omega(n))$ for any function $\omega(n)$ that tends to infinity. Details follow.

**Step 1**

Perform Steps 1 and 2 of the algorithm in section 5.1.

**step2**

*for* each $i$, $1 \leq i \leq (s+1)$ *do*

Choose $|X_i|/(\log\log n)^3$ keys at random from $X_i$. Sort these keys and pick keys in positions $1, (r'+1), (2r'+1), \ldots$ to form the splitter keys for this $X_i$ (where $r' = \log\log n$). Partition $X_i$ using these splitter keys and sort separately each resultant part.

**Analysis.** Step 1 of the above algorithm takes $O(n/(\log^2 n) + n(\log n - 4\log\log n))$ time.

Each $X_i$ will be of cardinality no more than $1.1 \log^4 n$ with high probability. Each $X_i$ can be sorted in time $|X_i| \log |X_i| + O(|X_i| \log\log |X_i|)$ with probability $\geq (1 - |X_i|^2 e^{-\log\log^2 n}) = (1 - \log^{-\Omega(1)} n)$. Thus, the expected time to sort $X_i$ is $|X_i| \log |X_i| + O(|X_i| \log\log |X_i|)$.

Summing over all $i$'s, the total expected time for Step 2 is

$$4n \log\log n + O(n) + O(n \log\log\log n).$$

Therefore, the expected run time of the whole algorithm is $n \log n + O(n \log\log\log n)$.

**Improvement:** The expected time bound of the above algorithm can be improved to $n \log n + O(n\ \omega(n))$. The idea is to employ more and more levels of recursion from Reischuk's algorithm.

# 6 Conclusions

In this paper we have derived randomized algorithms for selection and sorting. Many sampling lemmas have been proven which are most likely to find independent applications. For instance, lemma 2.2 has been used to design a constant time approximate median finding parallel algorithm on the CRCW PRAM [28].

# References

[1] A. Aho, J.E. Hopcroft, and J.D. Ullman, *The Design and Analysis of Algorithms*, Addison-Wesley Publications, 1976.

[2] M. Ajtai, J. Komlós, and E. Szemerédi, An $O(n \log n)$ Sorting Network, in Proc. ACM Symposium on Theory of Computing, 1983, pp. 1-9.

[3] D. Angluin and L.G. Valiant, Fast Probabilistic Algorithms for Hamiltonian Circuits and Matchings, Journal of Computer Systems and Science 18, 2, 1979, pp. 155-193.

[4] S. Carlsson, A Variant of Heapsort with Almost Optimal Number of Comparisons, Information Processing Letters 24, 1987, pp. 247-250.

[5] S. Carlsson, Average Case Results on Heapsort, BIT 27, 1987, pp. 2-17.

[6] H. Chernoff, A Measure of Asymptotic Efficiency for Tests of a Hypothesis Based on the Sum of Observations, Annals of Mathematical Statistics 23, 1952, pp. 493-507.

[7] R. Cole, Parallel Merge Sort, SIAM Journal on Computing, vol. 17, no. 4, 1988, pp. 770-785.

[8] R. Cole, An Optimally Efficient Selection Algorithm, Information Processing Letters 26, Jan. 1988, pp. 295-299.

[9] R. Cole and U. Vishkin, Approximate and Exact Parallel Scheduling with Applications to List, Tree, and Graph Problems, in Proc. IEEE Symposium on Foundations of Computer Science, 1986, pp. 478-491.

[10] W.D. Frazer and A.C. McKellar, Samplesort: A Sampling Approach to Minimal Storage Tree Sorting, Journal of the ACM, vol.17, no.3, 1970, pp. 496-507.

[11] R. Floyd and R. Rivest, Expected Time Bounds for Selection, Communications of the ACM, vol. 18, no. 3, 1975, pp. 165-172.

[12] C.A.R. Hoare, Quicksort, Computer Journal 5, 1962, pp. 10-15.

[13] C. Kaklamanis, D. Krizanc, L. Narayanan, and Th. Tsantilas, Randomized Sorting and Selection on Mesh Connected Processor Arrays, Proc. 3rd Annual ACM Symposium on Parallel Algorithms and Architectures, 1991.

[14] L. Kleinrock, *Queueing Theory. Volume 1: Theory*, John Wiley & Sons, 1975.

[15] D. Kozen, Semantics of Probabilistic Programs, Journal of Computer and Systems Science, vol. 22, 1981, pp. 328-350.

[16] D.E. Knuth, *The Art of Computer Programming*, vol. 3, Sorting and Searching, Addison-Wesley Publications, 1973.

[17] T. Leighton, Tight Bounds on the Complexity of Parallel Sorting, in Proc. ACM Symposium on Theory of Computing, 1984, pp. 71-80.

[18] F.P. Preparata, New Parallel Sorting Schemes, IEEE Transactions on Computers, vol. C27, no. 7, 1978, pp. 669-673.

[19] M.O. Rabin, Probabilistic Algorithms, in *Algorithms and Complexity, New Directions and Recent Results*, edited by J. Traub, Academic Press, 1976, pp. 21-36.

[20] S. Rajasekaran, $k-k$ Routing, $k-k$ Sorting, and Cut Through Routing on the Mesh, Technical Report MS-CIS-91-93, Department of CIS, University of Pennsylvania, October 1991. Also presented in the 4th Annual ACM Symposium on Parallel Algorithms and Architectures, 1992.

[21] S. Rajasekaran, Mesh Connected Computers with Fixed and Reconfigurable Buses: Packet Routing, Sorting, and Selection, Technical Report MS-CIS-92-56, Department of CIS, University of Pennsylvania, July 1992.

[22] S. Rajasekaran, Randomized Parallel Selection, Proc. Tenth Conference on Foundations of Software Technology and Theoretical Computer Science, Bangalore, India, 1990. Springer-Verlag Lecture Notes in Computer Science 472, pp. 215-224.

[23] S. Rajasekaran and J.H. Reif, Optimal and Sub Logarithmic Time Randomized Parallel Sorting Algorithms, SIAM Journal on Computing, vol. 18, no. 4, 1989, pp. 594-607.

[24] S. Rajasekaran and D.S.L. Wei, Selection, Routing, and Sorting on the Star Graph, to appear in Proc. 7th International Parallel Processing Symposium, 1993.

[25] J.H. Reif and L.G. Valiant, A Logarithmic Time Sort for Linear Size Networks, in Proc. 15th Annual ACM Symposium on Theory of Computing, Boston, MASS., 1983, pp. 10-16.

[26] J.H. Reif, An $n^{1+\epsilon}$ Processor $O(\log\log n)$ Time Probabilistic Sorting Algorithm, in Proc. SIAM Symposium on the Applications of Discrete Mathematics, Cambridge, MASS., 1983, pp. 27-29.

[27] R. Reischuk, Probabilistic Parallel Algorithms for Sorting and Selection, SIAM Journal on computing, vol. 14, 1985, pp. 396-409.

[28] S. Sen, Finding an Approximate Median with High Probability in Constant Parallel Time, Information Processing Letters 34, 1990, pp. 77-80.

[29] Y. Shiloach, and U. Vishkin, Finding the Maximum, Merging, and Sorting in a Parallel Computation Model, Journal of Algorithms 2, 1981, pp. 81-102.

[30] R. Solovay and V. Strassen, A Fast Monte-Carlo Test for Primality, SIAM Journal on Computing, vol. 6, 1977, pp. 84-85.

[31] L.G. Valiant, Parallelism in Comparison Problems, SIAM Journal on Computing, vol.4, 1975, pp. 348-355.

[32] I. Wegener, Bottom-up-Heapsort, a New Variant of Heapsort Beating of Average Quicksort (if $n$ is not very small), in Proc. Mathematical Foundations of Computer Science, Springer-Verlag Lecture Notes in Computer Science 452, 1990, pp. 516-522.

# Appendix: Chernoff Bounds for the Sum of Geometric Variables

A discrete random variable $X$ is said to be geometric with parameter $p$ if its probability mass function is given by $P[X = k] = q^{k-1}p$ (where $q = (1-p)$). $X$ can be thought of as the number of times a coin has to be flipped before a head appears, $p$ being the probability of getting a head in one flip.

Let $Y = \sum_{i=1}^{n} X_i$ where the $X_i$'s are independent and identically distributed geometric random variables with parameter $p$. ($Y$ can be thought of as the number of times a coin has to be flipped before a head appears for the $n$th time, $p$ being the probability that a head appears in a single flip).

In this section we are interested in obtaining probabilities in the tails of $Y$. Chernoff bounds introduced in [6] and later applied by Angluin and Valiant [3] is a powerful tool in computing such probabilities. (For a simple treatize on Chernoff bounds see [14, pp. 388-393]).

Let $M_X(v)$ and $M_Y(v)$ stand for the moment generating functions of $X$ and $Y$ respectively. Also, let $\Gamma_X(v) = \log M_X(v)$ and $\Gamma_Y(v) = \log M_Y(v)$. Clearly, $M_Y(v) = (M_X(v))^n$ and $\Gamma_Y(v) = n\Gamma_X(v)$.

The Chernoff bound for the tail of $Y$ is expressed as

$$P\left[Y \geq n\Gamma_X^{(1)}(v)\right] \leq exp\left(n\left[\Gamma_X(v) - v\Gamma_X^{(1)}(v)\right]\right)$$

for $v \geq 0$. In our case $M_X(v) = \frac{pe^v}{1-qe^v}$. $\Gamma_X(v) = \log p + v - \log(1-qe^v)$, and $\Gamma_X^{(1)} = \frac{1}{1-qe^v}$.

Thus the Chernoff bound becomes

$$P\left[Y \geq \frac{n}{1-qe^v}\right] \leq exp(n\left[\log p + v - \log(1-qe^v) - v/(1-qe^v)\right])$$

The RHS can be rewritten as

$$\left(\frac{pe^v}{1-qe^v}\right)^n exp\left(\frac{-vn}{1-qe^v}\right).$$

Substituting $(1+\epsilon)n/p$ for $n/(1-qe^v)$ we get,

$$P\left[Y \geq (1+\epsilon)\frac{n}{p}\right] \leq \left(\frac{q+\epsilon}{q}\right)^n \left[\frac{q(1+\epsilon)}{q+\epsilon}\right]^{\frac{(1+\epsilon)n}{p}}.$$

If $\epsilon << 1$, the above becomes

$$P\left[Y \geq (1+\epsilon)\frac{n}{p}\right] \leq exp\left(\frac{-\epsilon^2 n}{q}\right).$$

# 7

# Time-Space Optimal Parallel Computation

Michael A. Langston[1]
email: langston@cs.utk.edu
Department of Computer Science
University of Tennessee
Knoxville, TN 37996, USA

**Abstract**

The development of parallel file rearrangement algorithms that simultaneously optimize both time and space is surveyed. The classic problem of merging two sorted lists is used to illustrate fundamental techniques. Recent implementations on real parallel machines are also discussed. A primary aim of this research is to help narrow the gap between the theory and practice of parallel computing.

## 1 Introduction

The search for efficient nonnumerical parallel algorithms has been a long-standing topic of considerable interest. Foundational problems such as merging and sorting, as examples, have received enormous attention, as evidenced by the impressive volume of literature published on this subject (see [1, 6, 18] for recent surveys). Most of this quest has been for methods that are *time optimal* in the sense that they attain asymptotically optimal speedup[2]. Indeed, a number of parallel algorithms have been proposed that are optimal under this criterion, including those found in [2, 7, 5, 9, 17, 19, 21].

Unfortunately, however, little attention has been paid to pragmatic issues, most notably space utilization (see, for example, the formidable space management problems encountered when $\mathcal{NC}$-style algorithms[3] have been implemented on hypercube multiprocessors [4]). Despite the relatively low cost of memory today, space utilization continues to be a critical aspect in many applications, even for sequential processing; this criticality is only heightened in real parallel processing systems.

None of the algorithms referenced above is *time-space optimal*. That is, none achieves optimal speedup and, at the same time, requires only a constant amount of extra space per processor when the number of processors is fixed. New techniques change this picture. In [10] a parallel algorithm is described that, given an EREW PRAM[4] with $k$ processors, merges two sorted lists of total length $n$ in $O(n/k + \log n)$ time and $O(k)$ extra space. Thus this method is time-space optimal for any value of $k \leq n/(\log n)$. It naturally gives rise

[1]This research was partially supported by the National Science Foundation under grant MIP–8919312 and by the Office of Naval Research under contract N00014–90–J–1855.

[2]A parallel method attains asymptotically optimal speedup if the product of the number of processors it employs and the amount of time it takes is within a constant factor of the time required by a fastest sequential algorithm.

[3]A problem is said to be in $\mathcal{NC}$ if it possesses a parallel algorithm that, for any problem instance of size $n$, employs a number of processors bounded by some polynomial function of $n$ and requires an amount of time bounded by some polylogarithmic function of $n$.

[4]The EREW PRAM is the exclusive-read exclusive-write parallel random-access machine, a robust model of parallel computing. Results for this model automatically apply to more powerful models, such as the CREW (concurrent-read exclusive-write) PRAM.

to a time-space optimal sorting algorithm as well. In [11] time-space optimal algorithms are devised for the binary set and multiset operations. All these strategies can be made stable (preserving the original relative order of records with identical keys) with little additional effort.

The purpose of this chapter is to survey these new developments. In the next section, time-space optimal algorithms for the archetypical problem of merging are described. Other algorithms are outlined in Section 3, in an effort to illustrate the range of problems amenable to these methods. In Section 4, some computational experience gained to date on real parallel machines is discussed. A few concluding remarks are made in a final section.

# 2 A Sample Problem — Merging

## 2.1 A Brief Review of Sequential Merging

It is helpful first to review time-space optimal sequential merging. The optimality attained with respect to both time and space inherently relies on the related notions of *block rearranging* and *internal buffering*, ideas that can be traced back to [16]. A list containing $n$ records can be viewed as a collection of $\Theta(\sqrt{n})$ blocks, each of size $\Theta(\sqrt{n})$. Thus one block can be employed as an (internal) buffer to aid in resequencing the other blocks of the two sorted sublists and then merging these blocks into one sorted list. Since only the contents of the buffer and the relative order of the blocks need ever be out of sequence, linear time is sufficient to achieve order by straight-selection sorting [15] both the buffer and the blocks (each sort involves $O(\sqrt{n})$ keys). The interested reader is referred to [12]–[14] for extensive background, related results and additional details on these concepts.

For the sake of complete generality, neither the key nor any other part of a record may be modified. Such is necessary, for example, when records are write-protected or when there is no explicit key field within each record, but instead a record's key is a function of one or more of its data fields.

Let $L$ denote a list containing two sublists to be merged, each with its keys in nondecreasing order. A few simplifying assumptions are made about $L$ to facilitate discussion. (Implementation details for handling arbitrary lists are suppressed here, but can be found in [12].) It is assumed that $n$ is a perfect square, and that the records of $L$ have already been permuted so that $\sqrt{n}$ largest-keyed records are at the front of the list (their relative order there is immaterial), followed by the remainders of the two sublists, each of which is now assumed to contain an integral multiple of $\sqrt{n}$ records in nondecreasing order. Therefore, $L$ is viewed as a series of $\sqrt{n}$ blocks, each of size $\sqrt{n}$. The leading block will be used as an internal buffer to aid in the merge.

The first step is to sort the $\sqrt{n}-1$ rightmost blocks by their tails (rightmost elements), after which their tails form a nondecreasing key sequence. (In this setting, selection sort requires only $O(n)$ key comparisons and record exchanges.) Records within a block retain their original relative order.

The second step, which is the most complex, is to direct a sequence of series merges. An initial pair of series of records to be merged is located as follows. The first series begins with the head of block 2 and terminates with the tail of block $i$, $i \geq 2$, where block $i$ is the first block such that the key of the tail of block $i$ exceeds the key of the head of block $i+1$. The second series consists solely of the records of block $i+1$. The buffer is used to merge these two series. That is, the leftmost unmerged record in the first series is repeatedly compared to the leftmost unmerged record in the second, with the smaller-keyed record swapped with the leftmost buffer element. Ties are broken in favor of the leftmost series. (In general, the buffer may be broken into two pieces as the merge progresses.) This task is halted when the tail of block $i$ has been moved to its final position.

The next two series of records to be merged are now located. This time, the first begins with the leftmost unmerged record of block $i+1$ and terminates as before for some $j \geq i$. The second consists solely of the records of block $j+1$. The merge is resumed until the tail of block $j$ has been moved. This process of locating series of records and merging them is continued until a point is reached at which only one such series exists, which is merely shifted left, leaving the buffer in the last block.

The final step is to sort the buffer, thereby completing the merge of $L$. $O(n)$ time suffices for this entire procedure, because each step requires at most linear time. $O(1)$ space suffices as well, since the buffer was internal to the list, and since only a handful of additional pointers and counters are necessary.

## 2.2 Time-Space Optimal Parallel Merging

The sequential algorithm just described comprises three steps: *block sorting, series merging* and *buffer sorting*. Unfortunately, these steps do not appear to permit a direct parallelization, at least not one that requires only constant extra space per processor. In particular, the internal buffer is instrumental in the series merging step, dictating a block size of $\Theta(\sqrt{n})$ that in turn severely limits what can be accomplished efficiently in parallel.

Observe, however, that if a time-space optimal method were available that could use bigger blocks (one block of size of $n/k$ for each of the $k$ processors) and reorganize the file so that the problem is reduced to one of $k$ local merges, then a time-space optimal merge of $L$ could be completed by simply directing each processor to merge the contents of its own block using the algorithm sketched in the last section.

This observation is the genesis of the parallel method to be sketched. The algorithm comprises five steps: *block sorting, series delimiting, displacement computing, series splitting* and *local merging*. Since the last step (local merging) is easy from a parallel standpoint, it is perhaps not surprising that the earlier steps are relatively complicated.

To simplify the presentation, assume that the number of records in each of the two sublists in $L$ is evenly divisible by $k$. (Implementation details for handling arbitrary lists are omitted from this treatment, but can be found in [10].) A record or block from the first sublist of $L$ is referred to as an $L1$ record or an $L1$ block. The terms $L2$ record and $L2$ block are used in an analogous fashion for elements from the second sublist.

### 2.2.1 *Block Sorting*

$L$ is seen as a sequence of $k$ blocks, each of size $n/k$. The objective is to sort these blocks by their tails. This is a simple chore if one is willing to settle

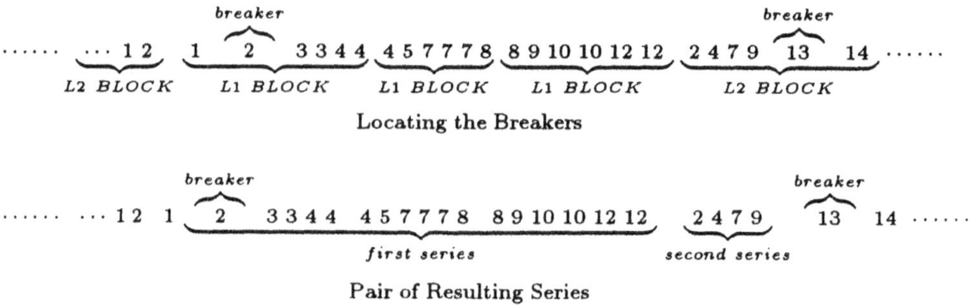


Figure 1: Delimiting a Pair of Series to be Merged

for a concurrent-read exclusive-write (CREW) algorithm. In order to sort the blocks efficiently on the EREW model, a slightly subtle strategy is needed. Each processor is first directed to set aside a copy of the tail of its block and its index (an integer between 1 and $k$, inclusive). The $k$ tail copies can now be merged (dragging along their indices) by reversing in parallel the copies from the second sublist and then invoking the well-known bitonic merge [3], a task requiring $O(\log k)$ time and $O(k)$ total extra space.

After this merge is completed, each processor knows the index of the block it is to receive. With the use of but one extra storage cell per processor, it is now a simple matter for the processors to acquire their respective new blocks in parallel without memory conflicts, one record at a time (say, from the first record in a block to the last). This task requires $O(n/k)$ time and $O(k)$ extra space.

### *2.2.2 Series Delimiting*

As with the sequential method, it is helpful at this point to think of the list as containing a collection of pairs of series of records, with each pair of series to be merged. The first and second series of any given pair meet as before, where the tail of block $i$ exceeds the head of block $i+1$. To determine where pairs meet each other, the term "breaker" is used to denote the first record of block $i+1$ that is no smaller than the tail of block $i$. Thus the first series of a pair needs only to begin with a breaker, and the second series of that pair needs only to end with the record immediately preceding the next breaker. This definition is illustrated in Fig. 1. Because each pair of series is made up either of a portion of an $L1$ block followed by zero or more full $L1$ blocks and a portion of an $L2$ block, or a portion of an $L2$ block followed by zero or more full $L2$ blocks and a portion of an $L1$ block, and because these two configurations are symmetric, only the former case is addressed in this and subsequent figures.

For a processor to determine whether its block contains a second series, it simply compares its head to its left neighbor's tail. If this comparison reveals that the processor does contain such a series, then it invokes a binary search to locate its breaker (it must have one — recall that the blocks were first sorted by their tails) and broadcasts[5] the breaker's location first to its left and then

[5] A convenient algorithm for this type of broadcasting can for example be found in [20],

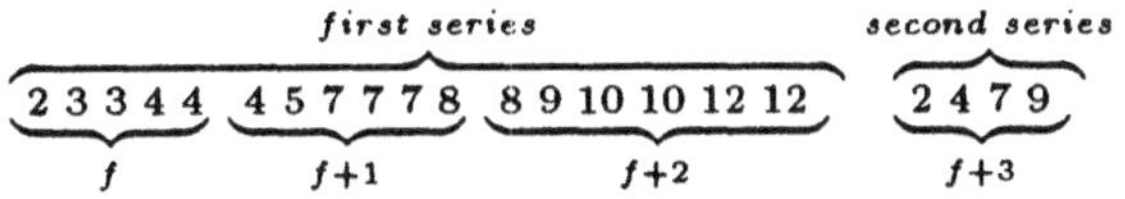


Pair of Series, p=3

| i | $E_i$ |
|---|---|
| f | 1 |
| f+1 | 2 |
| f+2 | 4 |

Displacement Table Entries

Figure 2: A Pair of Series and the Corresponding Displacement Table Entries

to its right. By this means, a processor learns the location of the breaker to its immediate right and the location of the breaker to its immediate left. From this it follows that every processor can correctly delimit the one or two pairs of series that are relevant to the contents of its block in $O(\log(n/k) + \log k)$ time and constant extra space per processor.

### *2.2.3 Displacement Computing*

A "displacement table" is now used, with one table entry to be stored at each processor. In this table is listed, for each processor with a block (or portion thereof) from the first series, the number of records from the second series that would displace records in that block *if* there were no other records in the first series. See Fig. 2.

Thus a displacement table is of immediate use in the next step (series splitting), because processor $i$ needs only to know its entry, $E_i$, and the entry for processor $i-1$, $E_{i-1}$. From these two values it is easy for processor $i$ to determine the number of its records that are to be displaced by records from the left (namely, $E_{i-1}$) and the number that are to be displaced by records from the second series (namely, $E_i - E_{i-1}$).

As with the block sorting step, things are relatively simple if one is willing to settle for a CREW algorithm. In order to compute the displacement table entries efficiently on the EREW model, a complicated strategy is employed. For an arbitrary pair of series, let $f$ denote the index of the processor handling the first record in the first series, and let $p$ denote the number of blocks with records in that series. Thus processor $f+p$ is responsible for the second series. The goal now is to direct the $p$ processors with records in the first series to work in unison and without memory conflicts to determine where each of their block's tails would need to go if they were merged with the $m < n/k$ records of the second series. To accomplish this, a technique is now presented that is perhaps best described as a sequence of phases of operations.

In the first phase, each processor with records in the first series sets aside a copy of its block's tail and its index (an integer between $f$ and $f+p-1$, inclusive). Each also sets aside two pieces of information from the second series: processor $i$ ($f \leq i < f+p$) computes and saves a copy of the offset

page 234, where it is termed a "data distribution algorithm." Alternately, such broadcasting can be efficiently accomplished with parallel prefix computation.

$h = (i - f + 1)(m/p)$ and a copy of the $h$th record of the second series. The $2p$ elements made up of $p$ tails and $p$ selected records (dragging along the indices and the offsets) can now be merged by reversing in parallel the selected records and then invoking a bitonic merge, a task requiring $O(\log p)$ time and $O(p)$ extra space.

After this, each processor with records in the first series examines the two keys in its temporary storage. If a processor finds a tail, then (with the use of the tail's index) it reports its own index to the processor handling the block from which the tail originated. Thus every processor can determine from the movement of its block's tail just how many of the records selected from the second series are smaller, and therefore which of the $p$ subseries of the second series, each subseries of size $m/p$, to merge into next. In order for a processor to be able to determine how many other tails are to be merged into the same next subseries as its block's tail, each one compares its next subseries with that of its neighbors. If the comparison reveals a subseries boundary, then broadcasting is used to inform the other processors of the location of this boundary (as done when broadcasting a breaker's location in the series delimiting step).

For the second and each subsequent phase, processors proceed as in the first phase, but now with new offsets and selected records based on the proper subseries into which their block's tails are to be merged and the number of other tails that are also to be merged there. Processors continue to iterate this procedure until each has determined where its block's tail would go if it were merged with the other tails and the second series. Note that some processors may be employed in as few as $\log_k m$ phases, each requiring $O(\log k)$ time, while others may simultaneously be employed in as many as $\log_2 m$ phases, each requiring constant time. In general, letting the sequence $k_1, k_2, ..., k_l$ denote the number of tails in any chain of recursive calls, observe that $k_1 \times k_2 \times ... \times k_l$ is $O(m)$, and hence $\log k_1 + \log k_2 + ... + \log k_l$ is $O(\log m)$. Therefore, $O(\log n)$ time and $O(k)$ extra space has been consumed up to this point.

Let $l_i$ $(1 \leq l_i \leq m + p)$ denote the location that the tail of the block of processor i $(f \leq i < f + p)$ would occupy in a sublist containing the $p$ tails and the entire second series if such a sublist were available. Processor $i$ now computes $l_i' = l_i - (i - f) - 1$, to eliminate the effect of its block's tail and all preceding tails. It next employs two pointers to compare a record in its block, beginning at location $n/k$ (its tail), to a record in the second series, beginning at location $l_i'$, repeatedly decrementing the pointer that points to the larger key for $l_i'$ iterations. (Each processor works from right to left in its interval of the second series in order to avoid memory conflicts. Processor $i$ keeps track of $l_{i-1}'$ and $l_{i+1}'$, relying on broadcasting by the leftmost processor if degeneracy in an interval occurs). When processor $i$ has finished decrementing its two pointers in this fashion, a task requiring $O(n/k)$ time and $O(k)$ extra space, the value of its second series pointer is its displacement table entry, $E_i$. Thus displacement computing can be accomplished in $O(n/k + \log n)$ time and constant extra space per processor.

### *2.2.4 Series Splitting*

At this point, processor $i$ can easily determine from the entries in the displacement table the number of its records that are to be displaced to the block to its right ($E_i$), as well as the number of records that it is to receive from the block

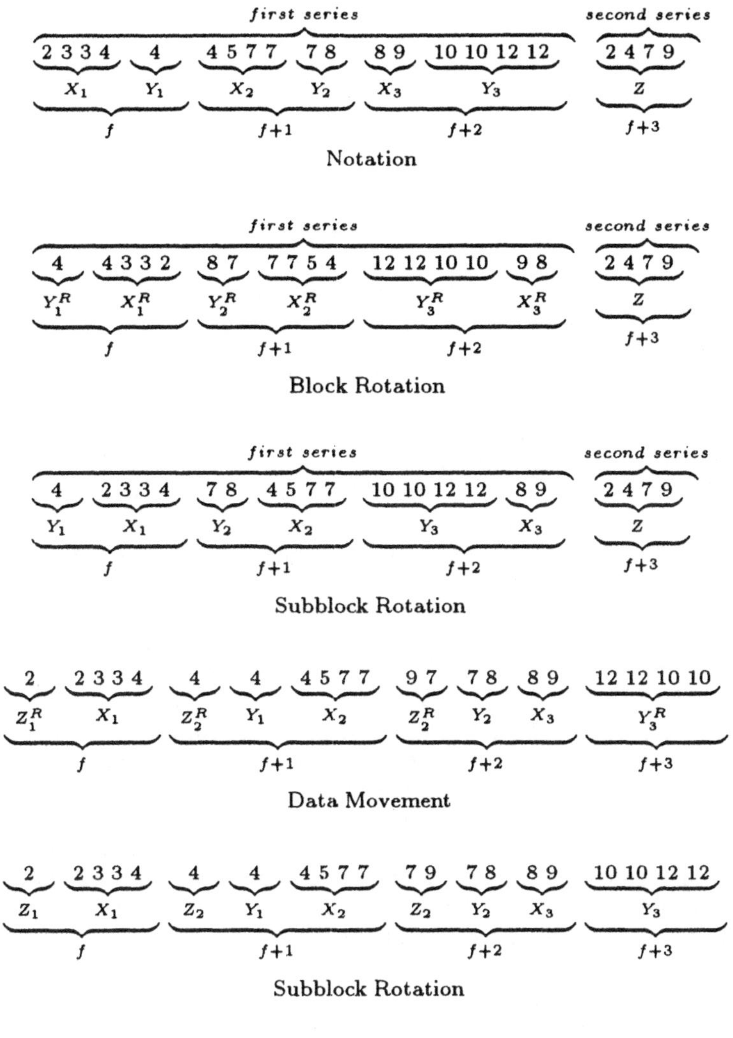


Figure 3: Series Splitting

to its left ($E_{i-1}$) and from the second series ($E_i - E_{i-1}$). Thus the second series is now split, in parallel, among the blocks of the first series. This is accomplished in constant extra space with the use of block rotations (each of which is effected with a sequence of three sublist reversals), followed by the desired data movement, followed by one last reversal. This procedure is illustrated in Fig. 3.

Letting $i$ denote the index of an arbitrary processor with records in the first series only, $X_i$ is used to denote its first $n/k - E_i$ records (that is, those to remain in this block) and $Y_i$ to denote the remaining $E_i$ records (that is, those to be displaced to the right). $Z$ is used to denote the contents of the portion of a block that constitutes the second series. Processor $i$ first reverses $X_i$ and $Y_i$ together, then each separately, thereby completing the rotation. Processor

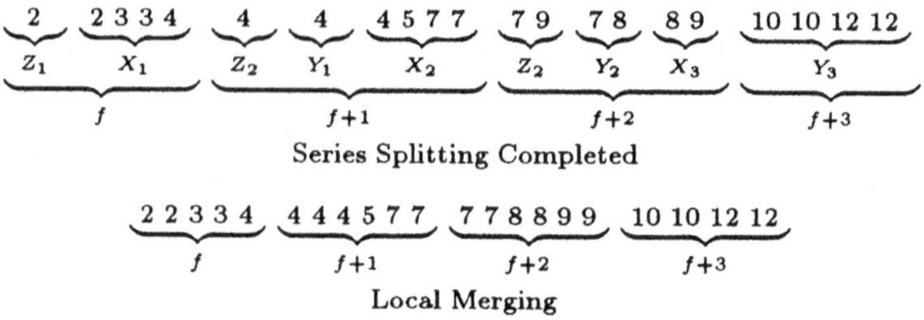


Figure 4: Local Merging

$i$ then initiates data movement, employing a single extra storage cell to copy safely the last record of $Y_i$ to the location formerly occupied by the last record of $Y_{i+1}$. (If processor $i$ is handling the last block of the first series, it instead copies its last $Y$ record to the former location of the first $Z$ record.) At the same time, the processor of the second series copies its first $Z$ record to the former location of the last $Y$ record of the first (portion of a) block in the first series. Continuing in this fashion, therefore, the data movement sequence is right-to-left for the blocks in the first series, but left-to-right for the second.

Of course, when block $i$ of the first series is filled, the processor of the second block must shift its attention to block $i+1$, and so on. If $k$ is small enough (no greater than $O(\log n)$), then the displacement table can simply be searched; if $k$ is larger than this, then the table may contain too many identical entries, and a preprocessing routine is invoked to condense it (again with the aid of broadcasting). The timing of the first and second series operations are interleaved (rather than simultaneous), because some processors will in general be handling portions of blocks of both types of series.

When the data movement phase is finished, each block will contain the correct prefix from the opposite series, but in reverse order. A final subblock reversal completes this step. Series splitting, therefore, requires $O(n/k+\log n)$ time and constant extra space per processor.

#### *2.2.5 Local Merging*

The linear-time, in-place sequential merge of the last subsection is employed. The completion of this merge is depicted in Fig. 4.

## 2.3 Merging Summary

In summary, the total time spent by the parallel merging algorithm is $O(n/k+\log n)$ and the total extra space used is $O(k)$. This method is therefore time-space optimal for any value of $k \leq n/(\log n)$. Moreover, it naturally provides a means for time-space optimal parallel sorting, providing improvements over the best previously-published PRAM methods designed for a bounded number of processors. For example, the recent EREW merging and sorting schemes proposed in [2] (where the issue of duplicate keys is not even addressed) are time optimal only for values of $k \leq n/(\log^2 n)$. More importantly, such schemes are not space optimal for *any* fixed $k$.

# 3 Other Amenable Problems

Just what scope of file rearrangement problems is amenable to time-space optimal parallel techniques? In this section, a partial answer to this question is provided by reviewing new time-space optimal parallel algorithms for the elementary binary set operations, namely, set union, intersection, difference and exclusive or. Most important is a handy procedure for selecting matched records.

## 3.1 Time-Space Optimal Parallel Selecting

Given two sorted lists $L1$ and $L2$, the goal is to transform $L1$ into two sorted sublists $L3$ and $L4$, where $L3$ consists of the records whose keys are not found in $L2$, and $L4$ consists of the records whose keys are. Thus $L = L1L2$ is input, and records are selected from $L1$ whose keys are contained in $L2$, accumulating them in $L4$, where the output is of the form $L3L4L2$.

The parallel algorithm comprises four steps: *local selecting, series delimiting, blockifying* and *block rearranging*. The number of records of each type ($L1$, $L2$, $L3$ and $L4$) is assumed to be evenly divisible by $k$, where $k$ denotes the number of processors available. (Implementation details are ignored here, but can be found in [11].)

### *3.1.1 Local Selecting*

$L$ is once again viewed as a collection of $k$ blocks, each of size $n/k$; a distinct processor is associated with each block. The idea is to treat each $L1$ block $L1_i$ as if it were the only block in $L1$, transforming its contents into the form $L3_iL4_i$.

The first task in this step is to determine where each tail (rightmost element) of each $L1$ block would go if the tails alone were to be merged with $L2$. In order to make this determination efficiently on the EREW model, each $L1$ processor is directed to set aside four extra storage cells (for copies of indices, offsets and keys) and to employ the "phased merge" as described in the displacement computing step of the parallel merge of the last section. At most $O(\log n)$ time and $O(k)$ extra space has been consumed up to this point.

As long as an $L1$ processor doesn't need to consider more than $O(n/k)$ $L2$ records (a quantity known by considering the difference between where its block's tail would go and where the tail of the block to its immediate left would go if they were to be merged with $L2$), it is instructed to employ the linear-time, in-place sequential select routine from [14]. Otherwise, in the case that an $L1$ block spans several $L2$ blocks, the corresponding $L2$ processors first preprocess their records (performing the time-space optimal sequential select against the $L1$ block, followed by a time-space optimal sequential duplicate-key extract [13]), then the $L1$ processor performs its select (at most $n/k$ $L2$ records are now needed), and finally the $L2$ processors restore their blocks (two time-space optimal sequential merge operations suffice).

Thus, letting $h$ denote the number of blocks in $L1$, the $L1$ list has now taken on the form $L3_1L4_1L3_2L4_2...L3_hL4_h$. This completes the local selecting step, and has required $O(n/k + \log n)$ time and constant extra space per processor.

$$\cdots \underbrace{L3_f^- L4_f \; L3_{f+1} L4_{f+1} \cdots \; L3_{g-1} L4_{g-1} \; L3_g L4_g \; L3_{g+1}^-}_{one\ series} \cdots$$

Figure 5: A Select Series

### *3.1.2 Series Delimiting*

$L1$ is now divided into a collection of non-overlapping series, each series with $n/k$ $L3$ records. This process is begun by locating breakers, each of which in this setting is the $(m(n/k)+1)$th $L3$ record for some integer $m$. Prefix sums are first computed on $|L3_i|$ to find these breakers. For example, if $\sum_{i=1}^{g-1} |L3_i| < m(n/k)+1$ and $\sum_{i=1}^{g} |L3_i| \geq m(n/k)+1$, then block $g$ contains the $m$th breaker. Three special types of breakers are identified. If block $i$ contains a breaker, but neither block $i-1$ nor block $i+1$ contain breakers, then the breaker in block $i$ is called a "lone" breaker. If block $i-1$ and block $i$ both contain breakers, and if block $i+1$ does not contain a breaker, then the breaker in block $i$ is called a "trailing" breaker. If block $i$ and block $i+1$ both contain breakers, and block $i-1$ does not contain a breaker, then the breaker in block $i$ is called a "leading" breaker.

These breakers are used to divide $L1$ into non-overlapping series as follows: each series begins with a lone or trailing breaker and ends with the record immediately preceding the next lone or leading breaker. By design, each series contains exactly $n/k$ $L3$ records. A sample series is depicted in Fig. 5, where $L3_f^-$ is used to denote $L3_f$ minus any records that precede its breaker and $L3_{g+1}^-$ to denote $L3_{g+1}$ minus its breaker and any records that follow it.

A processor that holds a lone or trailing breaker broadcasts its breaker's location to its right. After that, a processor that holds a lone or leading breaker broadcasts its breaker's location to its left. By this means, a processor learns the location of the lone or trailing breaker to its immediate left and the location of the lone or leading breaker to its immediate right. This completes the series delimiting step, and has required $O(\log(n/k)+\log k))$ time and constant extra space per processor.

### *3.1.3 Blockifying*

In this step, the $L1$ records within every series are first reorganized, then the records in the remainder of the $L1$ list are reorganized.

Reconsider the sample series. The goal is to collect the $n/k$ $L3$ records in this series in block $g$ (and thus move the $L4$ records into the other blocks and sub-blocks illustrated). It is a simple matter to exchange $L3_{g+1}^-$ with the rightmost $|L3_{g+1}^-|$ records in $L4_g$. Efficiently coalescing the other $L3$ records into block $g$ is much more difficult. Prefix sums on $|L3_f^-|$, $|L3_{f+1}|$, $\cdots$, $|L3_{g-2}|$, $|L3_{g-1}|$ are computed to obtain a displacement table. Table entry $E_i = \sum_{k=f}^{i} |L3_k|$ denotes the number of $L3$ records in blocks indexed $f$ through $i$ that are to move to block $g$. It turns out that $E_i$ will also denote the number of $L4$ records

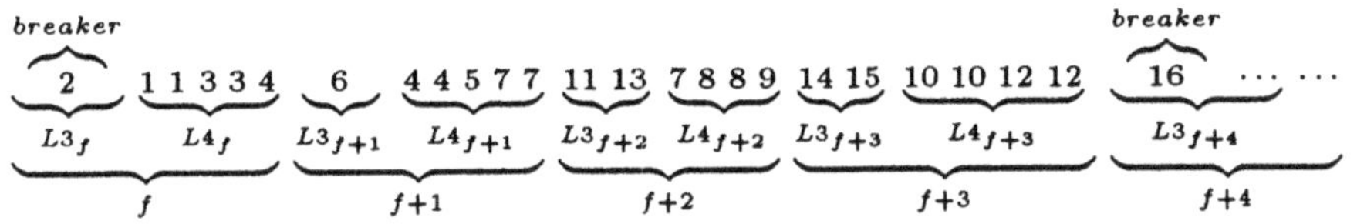


Series

| i | $E_i$ |
|---|---|
| f | 1 |
| f+1 | 2 |
| f+2 | 4 |

Displacement Table

Figure 6: A More Detailed View of a Select Series and its Displacement Table

that block $i$ is to receive from block $i+1$ as the algorithm proceeds. In Fig. 6, the sample series is shown in more detail (with $g$ set at $f+3$) along with its corresponding displacement table.

Thus each processor $i$, $f < i < g$, now uses the displacement table to determine exactly how the records in its block are to be rearranged: it is to send $|L3_i|$ records to block $g$, send its first $E_{i-1}$ $L4$ records (denoted by $X_i$) to block $i-1$, retain its next $n/k - |L3_i| - E_{i-1}$ $L4$ records (denoted by $Y_i$) and receive $E_i$ $L4$ records (denoted by $X_{i+1}$) from block $i+1$. Processors $f$ and $g$ determine similar information: processor $f$ is to send $|L3_f| = E_f$ records to block $g$ and receive the same number of records from block $f+1$; processor $g$ is to send $|L4_g| = E_{g-1}$ records to block $g-1$ and receive the same number of records from blocks $f$ through $g-1$. (Note that segments $X_f$ and $Y_g$ are empty.)

To accomplish the data movement, each processor first reverses the contents of its block, then reverses its $X$, $Y$ and $L3$ segments separately, thereby efficiently permuting its (two or) three subblocks. Each processor $j$, $f < j \leq g$, now employs a single extra storage cell to copy safely the first record of $X_j$ to the location formerly occupied by the first record of $X_{j-1}$, while processor $f$ copies the first record of its $L3$ segment to the location formerly occupied by the first record of $X_g$. Data movement continues in this fashion, with each processor moving its $L3$ records to block $g$ as soon as its $X$ segment is exhausted.

Note that if $k$ is small enough (no greater than $O(max\{n/k, \log n\})$), then the displacement table can merely be searched; if $k$ is larger than this, then the table may contain too many identical entries, and a preprocessing routine is invoked to condense it (again with the aid of broadcasting).

After the data movement is finished, it is necessary to rotate $L3_g$ with the records moved into block $g$ from block $g+1$. The processing of the series is now completed, as depicted in Fig 7.

If block $g+1$ contains a leading breaker, the records in an appropriate prefix of this block are rotated to ensure that $L3$ records precede $L4$ records there.

The records not spanned by a series can now be handled. These records are contained in zero or more non-overlapping "sequences" (a term chosen to avoid confusion with "series"), where each sequence begins with a leading breaker and ends with the record immediately preceding the next trailing breaker. Suppose

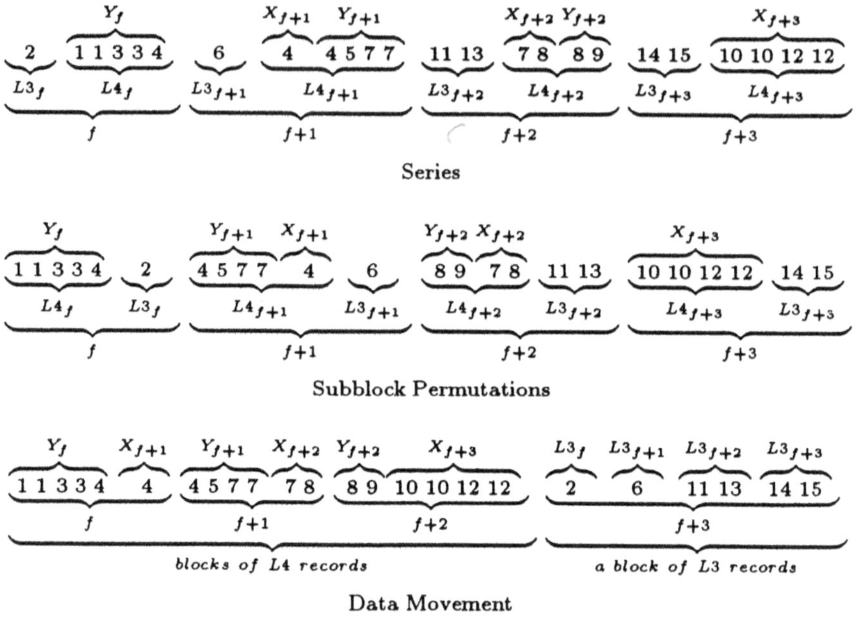


Figure 7: Coalescing the $L3$ Records into a Single Block

such a sequence spans $p$ blocks. Because there are exactly $p$ breakers in these blocks, and because the $L3$ records before the first breaker and after the last breaker have been moved outside these blocks, there are now exactly $(p-1)(n/k)$ $L3$ records there. Thus, there are exactly $n/k$ $L4$ records there.

If $p = 2$, then the two blocks have the form $L3_i L4_i L3_{i+1} L4_{i+1}$, where $|L4_i| = |L3_{i+1}|$. Swapping $L4_i$ with $L3_{i+1}$ finishes the blockifying for this sequence. If $p > 2$, then the sequence is treated as each series was earlier, exchanging the roles of $L3$ and $L4$ records. This completes the blockifying step, and has required $O(n/k + \log n)$ time and constant extra space per processor.

### *3.1.4 Block Rearranging*

$L3$ has now become an ordered collection of blocks interspersed with another ordered collection that constitutes $L4$. Now one needs only to rearrange these blocks so that $L3$ is followed by $L4$. Each processor is directed to set aside a zero-bit if it contains an $L3$ block, and to set aside a one-bit otherwise. The processors compute prefix sums on these values, and then acquire their respective new blocks in parallel without memory conflicts. This completes the block rearranging step, and has required $O(n/k + \log k)$ time and constant extra space per processor.

In summary, the total time spent by the parallel select algorithm is $O(n/k + \log n)$ and the total extra space used is $O(k)$. Therefore, like the merging routine of the last section, this algorithm is time-space optimal for any value of $k \leq n/(\log n)$.

### 3.2 Time-Space Optimal Parallel Set Operations

Consider the input list $L = XY$, where $X$ and $Y$ are two sublists, each sorted on the key, and each containing no duplicates. Three fundamental tools are sufficient: merge, select and duplicate-key extract. Merge and select have already been described. Duplicate-key extract is obtained from an easy modification to select, in which the first step, local selecting, is replaced with the local duplicate key extracting method of [13]. (Local duplicate-key extract is actually easier than local select, because the $L1$ processors need no information from the $L2$ list.)

Time-space optimal parallel routines for performing the elementary binary set operations are now at hand. Merge followed by duplicate-key extract produces $X \cup Y$. Select yields both $X \cap Y$ and $X - Y$. To achieve $X \oplus Y$, select is invoked on $XY$ producing $X_1X_2Y$, $X_2$ and $Y$ are rotated yielding $X_1YX_2$, select is invoked on $YX_2$ producing $X_1Y_1Y_2X_2$, and finally $X_1$ and $Y_1$ are merged.

As a bonus, these methods immediately extend to multisets (under several natural definitions [14]).

# 4 Practical Experience

Although the asymptotic optimality achieved is of interest from a theoretical perspective, experimental resources are now being employed to gauge the practical merit of these new methods. To make these algorithms effective on real machines, a number of difficulties must be overcome before any net run-time savings is realized. Notable difficulties include:

1) increased constants of proportionality (these routines are obviously quite complex, not mere parallelizations of sequential algorithms) and
2) synchronization overhead (frequently ignored in theory, synchronization can in practice quickly dominate *all* other computation and communication costs).

Representative results (for merging) are depicted in the next two figures. The values shown were obtained with a Sequent Symmetry with six processors, only five of which can be used by a single program. These experiments bear out that the methods discussed here attain linear speedup (see Fig. 8), despite several nontrivial implementation details.

Moreover, only four processors are needed to beat fast sequential analogs (see Fig. 9), which are vastly simpler and which incur no synchronization costs whatsoever.

These initial results are impressive, especially in light of the aforementioned difficulties associated with implementing PRAM-style algorithms. It is emphasized that the elapsed times illustrated count *everything*, including synchronization time (which is often conveniently omitted from experimental studies in the literature).

Large-scale implementations of these methods are now being conducted on a number of other MIMD and even SIMD machines[6]. Initial results appear very

[6] On an MIMD (multiple-instruction multiple-data) machine, processors may execute different programs on different data sets simultaneously. On an SIMD (single-instruction

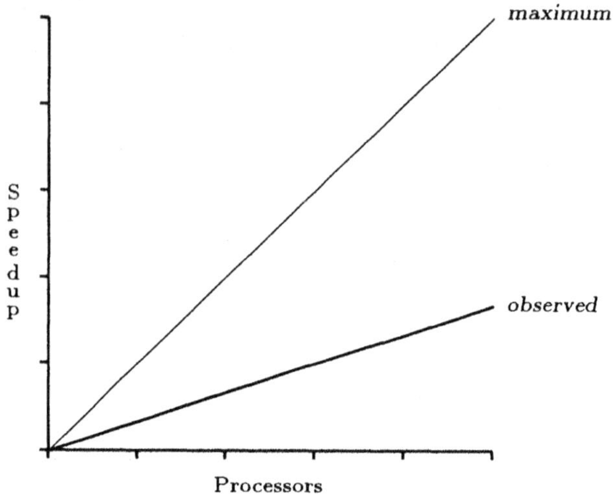


Figure 8: Observed Speedup of Parallel Algorithm

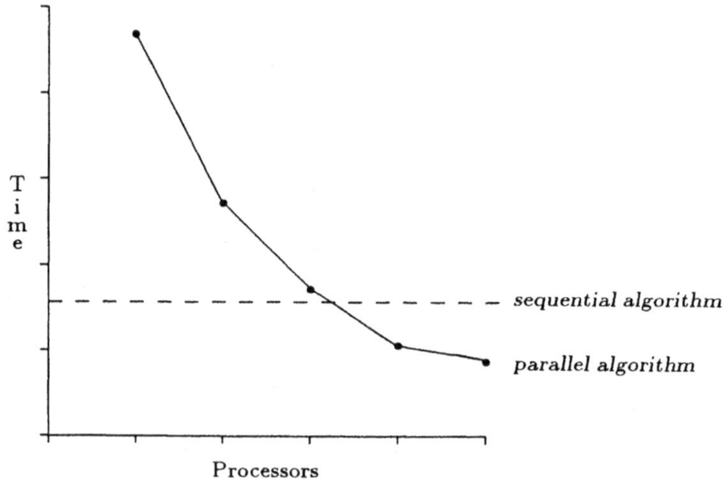


Figure 9: Comparison to Sequential Algorithm

promising for MIMD machines, even when memory is distributed among the processors rather than shared as it is on Sequents. The future is somewhat less certain for SIMD implementations[8]. Processors sometimes need to execute slightly different versions of a program. This can happen during merging, for example, when one processor receives an unusually short list or sublist. In this happenstance, versions must be run serially. (This sort of phenomenon is perhaps one reason for the apparent decline in the popularity of the SIMD model.)

# 5 Concluding Remarks

New parallel algorithms that are asymptotically time-space optimal have been surveyed. Remarkably, these methods assume only the weak EREW PRAM model. Although $n$ must be large enough so that the inequality $k \leq n/(\log n)$ is satisfied for optimality, these algorithms are efficient[7] for any value of $n$, suggesting that they may have practical merit even for relatively small inputs. For the sake of complete generality, these methods modify neither the key nor any other part of a record.

These algorithms are also *communication optimal* (assuming $k \leq n/(\log n)$). To see this, charge a data transfer to the sending processor and then count the number of messages sent by processor $i$, presupposing an input for which all of the data initially stored at processor $i$ must be transmitted elsewhere. Because only constant extra space is available at each processor, every message must be of constant length, and thus $\Omega(n/k)$ messages are charged to processor $i$ no matter the algorithm used. These methods are therefore optimal, since processor $i$ uses $O(n/k)$ time and hence sends at most $O(n/k)$ messages. (A similar argument holds if a data transfer is charged instead to the receiving processor.)

One might ask whether these methods can be improved to run in sub-logarithmic time. The answer is negative for merging on an EREW PRAM, because $\Omega(\log n)$ time is known to be a lower bound. Thus the parallel algorithm described in Section 2 is the best possible, to within a constant factor, for this model. (The situation is almost surely the same for the operations mentioned in Section 3.) Asymptotically faster time-space optimal algorithms may exist, however, for more powerful models. For example, it is an open question whether time-space optimal merging can be accomplished in $O(n/k + \log\log n)$ time on a CREW PRAM.

As long as memory management remains a critical aspect of many environments, the search for techniques that permit the efficient use of both time and space continues to be a worthwhile effort.

# References

[1] S. G. Akl, 'Parallel Sorting Algorithms', Academic Press, Orlando, FL, 1985.

multiple-data) machine, processors must execute the same program, though they may operate on different data sets.

[7] A parallel method is said to be efficient if its speedup is within a polylogarithmic factor of the optimum.

[2] S. G. Akl and N. Santoro, 'Optimal Parallel Merging and Sorting Without Memory Conflicts', *IEEE Transactions on Computers* 36 (1987), pp. 1367–1369.

[3] K. E. Batcher, 'Sorting Networks and their Application', *Proceedings, AFIPS 1968 Spring Joint Computer Conference* (1968), pp. 307–314.

[4] P. Banerjee and K. P. Belkhale, 'Parallel Algorithms for Geometric Connected Component Labeling Problems on a Hypercube', Technical Report, Coordinated Science Laboratory, University of Illinois, Urbana, IL, 1988.

[5] G. Baudet and D. Stevenson, 'Optimal Sorting Algorithms for Parallel Computers', *IEEE Transactions on Computers* 27 (1978), pp. 84–87.

[6] D. Bitton, D. J. Dewitt, D. K. Hsiao and J. Menon, 'A Taxonomy of Parallel Sorting', *Computing Surveys* 16 (1984), pp. 287–318.

[7] A. Borodin and J. E. Hopcroft, 'Routing, Merging and Sorting on Parallel Models of Computation', *Journal of Computer and System Sciences* 30 (1985), pp. 130–145.

[8] C. P. Breshears and M. A. Langston, 'MIMD versus SIMD Computation: Experience with Non-Numeric Parallel Algorithms', in *Proc. Twenty-Sixth Annual Hawaii International Conference on System Sciences, Vol. II: Software Technology*, H. El-Rewini, T. Lewis, and B. D. Shriver (editors), 1993, pp. 298–307.

[9] R. Cole, 'Parallel Merge Sort', *SIAM Journal on Computing* 17 (1988), pp. 770–785.

[10] X. Guan and M. A. Langston, 'Time-Space Optimal Parallel Merging and Sorting', *IEEE Transactions on Computers* 40 (1991), pp. 596–602.

[11] X. Guan and M. A. Langston, 'Parallel Methods for Solving Fundamental File Rearrangement Problems', *Journal of Parallel and Distributed Computing* 14, (1992), pp. 436–439.

[12] B-C Huang and M. A. Langston, 'Practical In-Place Merging', *Communications of the ACM* 31 (1988), pp. 348–352.

[13] B-C Huang and M. A. Langston, 'Stable Duplicate-Key Extraction with Optimal Time and Space Bounds', *Acta Informatica* 26 (1989), pp. 473–484.

[14] B-C Huang and M. A. Langston, 'Stable Set and Multiset Operations in Optimal Time and Space', *Information Processing Letters* 39 (1991), pp. 131–136.

[15] D. E. Knuth, 'The Art of Computer Programming, Vol. 3: Sorting and Searching', Addison-Wesley, Reading, MA, 1973.

[16] M. A. Kronrod, 'An Optimal Ordering Algorithm without a Field of Operation', *Doklady Akademii Nauk SSSR* 186 (1969), pp. 1256–1258.

[17] C. P. Kruskal, 'Searching, Merging and Sorting in Parallel Computation', *IEEE Transactions on Computers* 32 (1983), pp. 942–946.

[18] S. Lakshmivarahan, S. K. Dhall, and L. L. Miller, 'Parallel Sorting Algorithms', *Advances in Computers* 23 (1984), pp. 295–354.

[19] Y. Shiloach and U. Vishkin, 'Finding the Maximum, Merging and Sorting in a Parallel Computation Model', *Journal of Algorithms* 2 (1981), pp. 88–102.

[20] J. D. Ullman, 'Computational Aspects of VLSI', Computer Science Press, Rockville, MD, 1984.

[21] L. G. Valiant, 'Parallelism in Comparison Problems', *SIAM Journal on Computing* 4 (1975), pp. 349–355.

# INDEX

**P**



**R**



**S**

GPSR Compliance

*The European Union's (EU) General Product Safety Regulation (GPSR) is a set of rules that requires consumer products to be safe and our obligations to ensure this.*

*If you have any concerns about our products, you can contact us on ProductSafety@springernature.com*

In case Publisher is established outside the EU, the EU authorized representative is:

Springer Nature Customer Service Center GmbH
Europaplatz 3
69115 Heidelberg, Germany

**Batch number: 10761599**

Printed by Printforce, the Netherlands